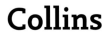

2018 GUIDE
to the
NIGHT SKY

Storm Dunlop and Wil Tirion

Published by Collins
An imprint of HarperCollins Publishers
Westerhill Road
Bishopbriggs
Glasgow G64 2QT
www.harpercollins.co.uk

In association with
Royal Museums Greenwich, the group name for the National Maritime Museum,
Royal Observatory Greenwich, Queen's House and *Cutty Sark* 2017
www.rmg.co.uk

A catalogue record for this book is available from the British Library

ISBN 978-0-00-824947-2

10 9 8 7 6 5 4 3 2 1

Printed in China by RR Donnelley APS

If you would like to comment on any aspect of this book, please contact us at the above address or online.
e-mail: collinsmaps@harpercollins.co.uk

 facebook.com/CollinsAstronomy

@CollinsAstro

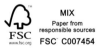

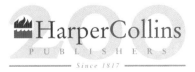

HarperCollins
PUBLISHERS
Since 1817

Contents

Introduction

The aim of this Guide is to help people find their way around the night sky, by showing how the stars that are visible change from month to month and by including details of various events that occur throughout the year. The objects and events described may be observed with the naked eye, or nothing more complicated than a pair of binoculars.

The conditions for observing naturally vary over the course of the year. During the summer, twilight may persist throughout the night and make it difficult to see the faintest stars. There are three recognized stages of twilight: civil twilight, when the Sun is less than 6° below the horizon; nautical twilight, when the Sun is between 6° and 12° below the horizon; and astronomical twilight, when the Sun is between 12° and 18° below the horizon. Full darkness occurs only when the Sun is more than 18° below the horizon. During nautical twilight, only the very brightest stars are visible. During astronomical twilight, the faintest stars visible to the naked eye may be seen directly overhead, but are lost at lower altitudes. As the diagram shows, during June and most of July full darkness never occurs at the latitude of London, and at Edinburgh

nautical twilight persists throughout the whole night, so at that latitude only the very brightest stars are visible.

Another factor that affects the visibility of objects is the amount of moonlight in the sky. At Full Moon, it may be very difficult to see some of the fainter stars and objects, and even when the Moon is at a smaller phase it may seriously interfere with visibility if it is near the stars or planets in which you are interested. A full lunar calendar is given for each month and may be used to see when nights are likely to be darkest and best for observation.

The celestial sphere

All the objects in the sky (including the Sun, Moon and stars) appear to lie at some indeterminate distance on a large sphere, centred on the Earth. This *celestial sphere* has various reference points and features that are related to those of the Earth. If the Earth's rotational axis is extended, for example, it points to the North and South Celestial Poles, which are thus in line with the North and South Poles on Earth. Similarly, the *celestial equator* lies in the same plane as the Earth's equator, and divides the sky into northern and

The duration of twilight throughout the year at London and Edinburgh.

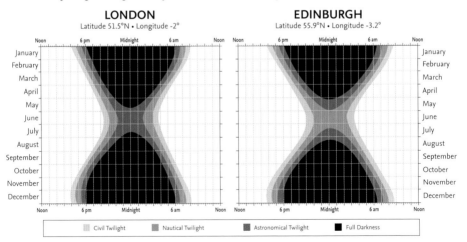

southern hemispheres. Because this Guide is written for use in Britain and Ireland, the area of the sky that it describes includes the whole of the northern celestial hemisphere and those portions of the southern that become visible at different times of the year. Stars in the far south, however, remain invisible throughout the year, and are not included.

It is useful to know some of the special terms for various parts of the sky. As seen by an observer, half of the celestial sphere is invisible, below the horizon. The point directly overhead is known as the **zenith**, and the (invisible) one below one's feet as the **nadir**. The line running from the north point on the horizon, up through the zenith and then down to the south point is the **meridian**. This is an important invisible line in the sky, because objects are highest in the sky, and thus easiest to see, when they cross the meridian in the south. Objects are said to **transit**, when they cross this line in the sky.

In this book, reference is frequently made in the text and in the diagrams to the standard compass points around the horizon. The position of any object in the sky may be described by its **altitude** (measured in degrees above the horizon), and its **azimuth** (measured

in degrees from north 0°, through east 90°, south 180° and west 270°). Experienced amateurs and professional astronomers also use another system of specifying locations on the celestial sphere, but that need not concern us here, where the simpler method will suffice.

The celestial sphere appears to rotate about an invisible axis, running between the North and South Celestial Poles. The location (i.e., the altitude) of the Celestial Poles depends entirely on the observer's position on Earth or, more specifically, their latitude. The charts in this book are produced for the latitude of 50°N, so the North Celestial Pole (NCP) is 50° above the northern horizon. The fact that the NCP is fixed relative to the horizon means that all stars within 50° of the pole are always above the horizon and may, therefore, always be seen at night, regardless of the time of year. This northern **circumpolar region** is an ideal place to begin learning the sky, and ways to identify the circumpolar stars and constellations will be described shortly.

The ecliptic and the zodiac
Another important line on the celestial sphere is the Sun's apparent path against the background stars – in reality the result

Measuring altitude and azimuth on the celestial sphere.

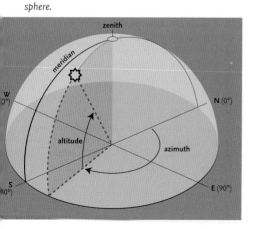

The altitude of the North Celestial Pole equals the observer's latitude.

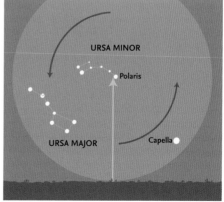

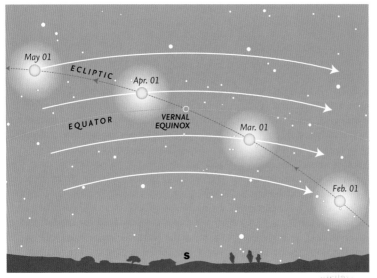

The Sun crossing the celestial equator in spring.

of the Earth's orbit around the Sun. This is known as the **ecliptic**. The point where the Sun, apparently moving along the ecliptic, crosses the celestial equator from south to north is known as the vernal (or spring) equinox, which occurs on March 20. At this time (and at the autumnal equinox, on September 22 or 23, when the Sun crosses the celestial equator from north to south) day and night are almost exactly equal in length. (There is a slight difference, but that need not concern us here.) The vernal equinox is currently located in the constellation of Pisces, and is important in astronomy because it defines the zero point for a system of celestial coordinates, which is, however, not used in this Guide.

The Moon and planets are to be found in a band of sky that extends 8° on either side of the ecliptic. This is because the orbits of the Moon and planets are inclined at various angles to the ecliptic (i.e., to the plane of the Earth's orbit). This band of sky is known as the Zodiac and, when originally devised, consisted of twelve **constellations**, all of which were considered to be exactly 30° wide. When the constellation boundaries were formally established by the International Astronomical

Union in 1930, the exact extent of most constellations was altered and, nowadays, the ecliptic passes through thirteen constellations. Because of the boundary changes, the Moon and planets may actually pass through several other constellations that are adjacent to the original twelve.

The constellations

Since ancient times, the celestial sphere has been divided into various constellations, most dating back to antiquity and usually associated with certain myths or legendary people and animals. Nowadays, the boundaries of the constellations have been fixed by international agreement and their names (in Latin) are largely derived from Greek or Roman originals. Some of the names of the most prominent stars are of Greek or Roman origin, but many are derived from Arabic names. Many bright stars have no individual names and, for many years, stars were identified by terms such as 'the star in Hercules' right foot'. A more sensible scheme was introduced by the German astronomer Johannes Bayer in the early seventeenth century. Following his scheme – which is still used today – most of the brightest stars are identified by a Greek letter followed by the genitive form of the constellation's Latin name. An example is the Pole Star, also known as Polaris and α Ursae Minoris (abbreviated α UMi). The Greek alphabet is shown on page 93 with a list of all the constellations that may be seen from latitude 50°N, together with abbreviations, their genitive forms and English names. Other naming schemes exist for fainter stars, but are not used in this book.

Asterisms

Apart from the constellations (88 of which cover the whole sky), certain groups of stars, which may form a part of a larger constellation or cross several constellations, are readily recognizable and have been given individual names. These groups are known as **asterisms**, and the most famous (and well-known) is the 'Plough', the common name for the seven brightest stars in the constellation of Ursa Major, the Great Bear. The names and details of some asterisms mentioned in this book are given in the list on page 94.

Magnitudes

The brightness of a star, planet or other body is frequently given in magnitudes (mag.). This is a mathematically defined scale where larger numbers indicate a fainter object. The scale extends beyond the zero point to negative numbers for very bright objects. (Sirius, the brightest star in the sky is mag. -1.4.) Most observers are able to see stars to about mag. 6, under very clear skies.

The Moon

Although the daily rotation of the Earth carries the sky from east to west, the Moon gradually moves eastwards by approximately its diameter (about half a degree) in an hour. Normally, in its orbit around the Earth, the Moon passes above or below the direct line between Earth and Sun (at New Moon) or outside the area obscured by the Earth's shadow (at Full Moon).

Occasionally, however, the three bodies are more-or-less perfectly aligned to give an **eclipse**: a solar eclipse at New Moon or a lunar eclipse at Full Moon. Depending on the exact circumstances, a solar eclipse may be merely partial (when the Moon does not cover the whole of the Sun's disk); annular (when the Moon is too far from Earth in its orbit to appear large enough to hide the whole of the Sun); or total. Total and annular eclipses are visible from very restricted areas of the Earth, but partial eclipses are normally visible over a wider area.

Somewhat similarly, at a lunar eclipse, the Moon may pass through the outer zone of the Earth's shadow, the **penumbra** (in a penumbral eclipse, which is not generally perceptible to the naked eye); so that just part of the Moon is within the darkest part of the Earth's shadow, the **umbra** (in a partial eclipse); or completely within the umbra (in a total eclipse). Unlike solar eclipses, lunar eclipses are visible from large areas of the Earth.

Occasionally, as it moves across the sky, the Moon passes between the Earth and individual planets or distant stars, giving rise to an **occultation**. As with solar eclipses, such occultations are visible from restricted areas of the world.

The planets

Because the planets are always moving against the background stars, they are treated in some detail in the monthly pages and information is given when they are close to other planets, the Moon or any of five bright stars that lie near the ecliptic. Such events are known as **appulses** or, more frequently, as **conjunctions**. (There are technical differences in the way these terms are defined – and should be used – in astronomy, but these need not concern us here.) The positions of the planets are shown for every month on a special chart of the ecliptic.

The term conjunction is also used when a planet is either directly behind or in front of

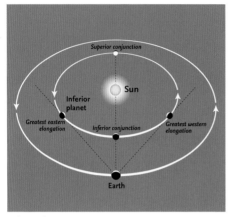

Inferior planet.

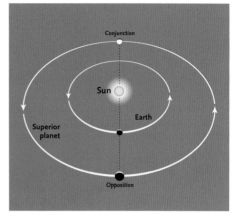

Superior planet.

the Sun, as seen from Earth. (Under normal circumstances it will then be invisible.) The conditions of most favourable visibility depend on whether the planet is one of the two known as **inferior planets** (Mercury and Venus) or one of the three **superior planets** (Mars, Jupiter and Saturn) that are covered in detail. (Some details of the fainter superior planets, Neptune and Uranus, are included in this Guide, and special charts for both are given on pages 71 and 75.)

The inferior planets are most readily seen at eastern or western **elongation**, when their angular distance from the Sun is greatest. For superior planets, they are best seen at **opposition**, when they are directly opposite the Sun in the sky, and cross the meridian at local midnight.

It is often useful to be able to estimate angles on the sky, and approximate values may be obtained by holding one hand at arm's length. The various angles are shown in the diagram, together with the separations of the various stars in the Plough.

Meteors

At some time or other, nearly everyone has seen a **meteor** – a 'shooting star' – as it flashed across the sky. The particles that cause meteors – known technically as 'meteoroids' – range in size from that of a grain of sand (or even smaller) to the size of a pea. On any night of the year there are occasional meteors, known as **sporadics**, that may travel in any direction. These occur at a rate that is normally between three and eight in an hour. Far more important, however, are **meteor showers**, which occur at fixed periods of the year, when the Earth encounters a trail of particles left behind by a comet or, very occasionally, by a minor planet (asteroid). Meteors always appear to diverge from a single point on the sky, known as the **radiant**, and the radiants of major showers are shown on the charts. Meteors that come from a circular area 8° in diameter around the radiant are classed as belonging to the particular shower. All others that do not come from that area are sporadics (or, occasionally from another shower that is active at the same time). A list of the major meteor showers is given on page 17.

Although the positions of the various shower radiants are shown on the charts, looking directly at the radiant is not the most effective way of seeing meteors. They are most likely to be noticed if one is looking about 40–45° away from the radiant position. (This is approximately two hand-spans as shown in the diagram for measuring angles.)

Other objects

Certain other objects may be seen with the naked eye under good conditions. Some were given names in antiquity – Praesepe is one

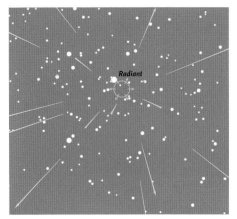

Meteor shower (showing the April Lyrid radiant).

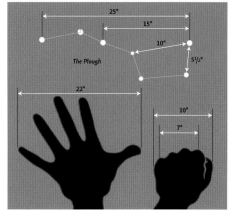

Measuring angles in the sky.

example – but many are known by what are called 'Messier numbers', the numbers in a catalogue of nebulous objects compiled by Charles Messier in the late eighteenth century. Some, such as the Andromeda Galaxy, M31, and the Orion Nebula, M42, may be seen by the naked eye, but all those given in the list will benefit from the use of binoculars. Apart from galaxies, such as M31, which contain thousands of millions of stars, there are also two types of cluster: open clusters,

such as M45, the Pleiades, which may consist of a few dozen to some hundreds of stars; and globular clusters, such as M13 in Hercules, which are spherical concentrations of many thousands of stars. One or two gaseous nebulae, consisting of gas illuminated by stars within them, are also visible. The Orion Nebula, M42, is one, and is illuminated by the group of four stars, known as the Trapezium, which may be seen within it by using a good pair of binoculars.

Some interesting objects.

Messier / NGC	Name	Type	Constellation	Maps (months)
—	Hyades	open cluster	Taurus	Sep. – Apr.
—	Double Cluster	open cluster	Perseus	All year
—	Melotte 111 (Coma Cluster)	open cluster	Coma Berenices	Jan. – Aug.
M3	—	globular cluster	Canes Venatici	Jan. – Sep.
M4	—	globular cluster	Scorpius	May – Aug.
M8	Lagoon Nebula	gaseous nebula	Sagittarius	Jun. – Sep.
M11	Wild Duck Cluster	open cluster	Scutum	May – Oct.
M13	Hercules Cluster	globular cluster	Hercules	Feb. – Nov.
M15	—	globular cluster	Pegasus	Jun. – Dec.
M22	—	globular cluster	Sagittarius	Jun. – Sep.
M27	Dumbbell Nebula	planetary nebula	Vulpecula	May – Dec.
M31	Andromeda Galaxy	galaxy	Andromeda	All year
M35	—	open cluster	Gemini	Oct. – May
M42	Orion Nebula	gaseous nebula	Orion	Nov. – Mar.
M44	Praesepe	open cluster	Cancer	Nov. – Jun.
M45	Pleiades	open cluster	Taurus	Aug. – Apr.
M57	Ring Nebula	planetary nebula	Lyra	Apr. – Dec.
M67	—	open cluster	Cancer	Dec. – May
NGC 752	—	open cluster	Andromeda	Jul. – Mar.
NGC 3242	Ghost of Jupiter	planetary nebula	Hydra	Feb. – May

The Northern Circumpolar Constellations

The northern circumpolar stars are the key to starting to identify the constellations. For anyone in the northern hemisphere they are visible at any time of the year, and nearly everyone is familiar with the seven stars of the Plough – known as the Big Dipper in North America – an asterism that forms part of the large constellation of Ursa Major (the Great Bear).

Ursa Major

Because of the movement of the stars caused by the passage of the seasons, Ursa Major lies in different parts of the evening sky at different periods of the year. The diagram below shows its position for the four main seasons. The seven stars of the Plough remain visible throughout the year anywhere north of latitude 40°N. Even at the latitude (50°N) for which the charts in this book are drawn, many of the stars in the southern portion of the constellation of Ursa Major are hidden below the horizon for part of the year or (particularly in late summer) cannot be seen late in the night.

Polaris and Ursa Minor

The two stars **Dubhe** and **Merak** (α and β Ursae

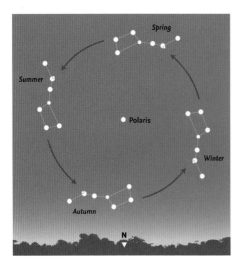

Majoris, respectively), farthest from the 'tail' are known as the 'Pointers'. A line from Merak to Dubhe, extended about five times their separation, leads to the Pole Star, **Polaris**, or α Ursae Minoris. All the stars in the northern sky appear to rotate around it. There are five main stars in the constellation of **Ursa Minor**, and the two farthest from the Pole, **Kochab** and **Pherkad** (β and γ Ursae Minoris, respectively), are known as 'The Guards'.

Cassiopeia

On the opposite of the North Pole from Ursa Major lies **Cassiopeia**. It is highly distinctive, appearing as five stars forming a letter 'W' or 'M' depending on its orientation. Provided the sky is reasonably clear of clouds, you will nearly always be able to see either Ursa Major or Cassiopeia, and thus be able to orientate yourself on the sky.

To find Cassiopeia, start with **Alioth** (ε Ursae Majoris), the first star in the tail of the Great Bear. A line from this star extended through Polaris points directly towards γ Cassiopeiae, the central star of the five.

Cepheus

Although the constellation of **Cepheus** is fully circumpolar, it is not nearly as well-known as Ursa Major, Ursa Minor or Cassiopeia, partly because its stars are fainter. Its shape is rather like the gable end of a house. The line from the Pointers through Polaris, if extended, leads to **Errai** (γ Cephei) at the 'top' of the 'gable'. The brightest star, **Alderamin** (α Cephei) lies in the Milky Way region, at the 'bottom right-hand corner' of the figure.

Draco

The constellation of **Draco** consists of a quadrilateral of stars, known as the 'Head of Draco' (and also the 'Lozenge'), and a long chain of stars forming the neck and body of the dragon. To find the Head of Draco, locate the two stars **Phecda** and **Megrez** (γ and δ Ursae Majoris) in the Plough, opposite the Pointers.

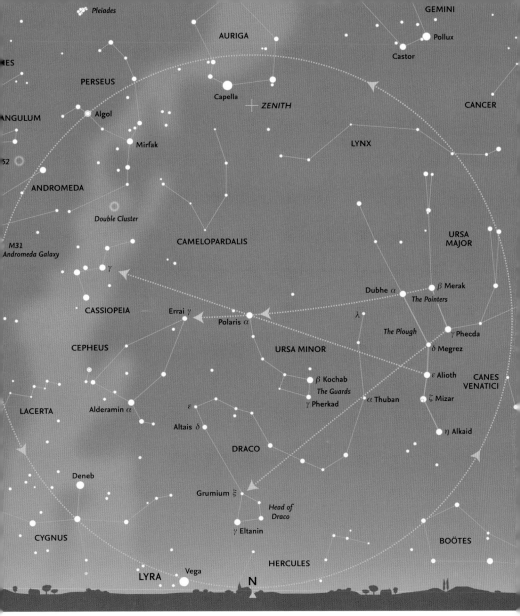

The stars and constellations inside the circle are always above the horizon, seen from our latitude.

Extend a line from Phecda through Megrez by about eight times their separation, right across the sky below the Guards in Ursa Minor, ending at **Grumium** (ξ Draconis) at one corner of the quadrilateral. The brightest star, **Eltanin** (γ Draconis) lies farther to the south. From the head of Draco, the constellation first runs northeast to **Altais** (δ Draconis) and ε Draconis, then doubles back southwards before winding its way through **Thuban** (α Draconis) before ending at λ Draconis between the Pointers and Polaris.

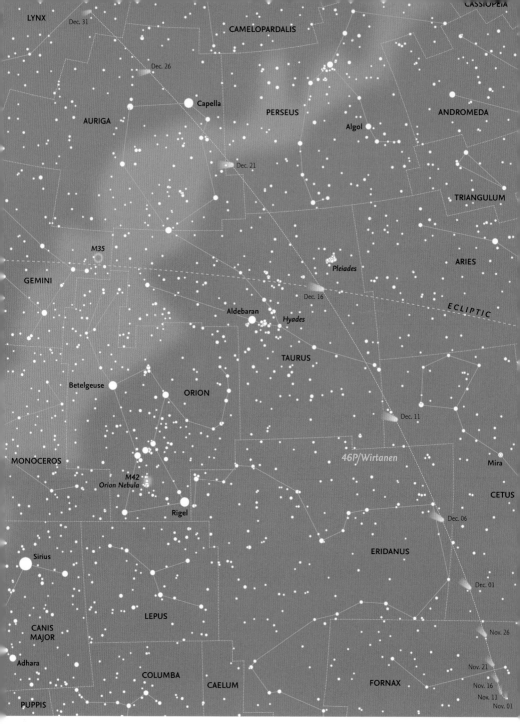

The path of comet 46P/Wirtanen from November 1 to December 31, 2018. Stars down to magnitude 6.5 are shown.

Comets and the Moon

Comets

Although comets may occasionally become very striking objects in the sky, their occurrence and particularly the existence or length of any tail and their overall magnitude are notoriously difficult to predict. Naturally, it is only possible to predict the return of periodic comets (whose names have the prefix 'P'). Many comets appear unexpectedly (these have names with the prefix 'C'). Bright, readily visible comets such as C/1995 Y1 Hyakutake & C/1995 O1 Hale-Bopp are rare. (The latter, in particular, was visible for a record 18 months and was a prominent object in northern skies.) Most periodic comets are faint and only a very small number ever become bright enough to be readily visible with the naked eye or with binoculars. In 2018, one comet, 46P/Wirtanen, is expected to become visible in binoculars in October, and may even become visible to the naked eye in December when it is well placed in the northern sky. (It should also be visible in the first few months of 2019.) The accompanying chart shows its path when it is expected to be brightest during November and December 2018.

Comet C/2014 Q2 Lovejoy, which reached naked-eye visibility, photographed on 20 December 2014, when close to κ Columbae, by Damian Peach.

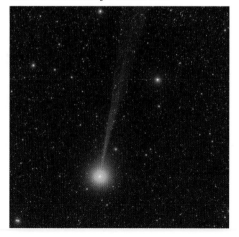

The Moon

The monthly pages include diagrams showing the phase of the Moon for every day of the month, and also indicate the day in the **lunation** (or *age* of the Moon), which begins at New Moon. Although the main features of the surface – the light highlands and the dark maria (seas) – may be seen with the naked eye, far more features may be detected with the use of binoculars or any telescope. The many craters are best seen when they are close to the **terminator** (the boundary between the illuminated and the non-illuminated areas of the surface), when the Sun rises or sets over any particular region of the Moon and the crater walls or central peaks cast strong shadows. Most features become difficult to see at Full Moon, although this is the best time to see the bright ray systems surrounding certain craters. Accompanying the Moon map on the following pages is a list of prominent features, including the days in the lunation when features are normally close to the terminator and thus easiest to see. A few bright features such as Linné and Proclus, visible when well illuminated, are also listed. One feature, Rupes Recta (the Straight Wall) is readily visible only when it casts a shadow with light from the east, appearing as a light line when illuminated from the opposite direction.

The dates of visibility vary slightly through the effects of **libration**. Because the Moon's orbit is inclined to the Earth's equator and also because it moves in an ellipse, the Moon appears to rock slightly from side to side (and nod up and down). Features near the **limb** (the edge of the Moon) may vary considerably in their location and visibility. (This is easily noticeable with Mare Crisium and the craters Tycho and Plato.) Another effect is that at crescent phases before and after New Moon, the normally non-illuminated portion of the Moon receives a certain amount of light, reflected from the Earth. This **Earthshine** may enable certain bright features (such as Aristarchus, Kepler and Copernicus) to be detected even though they are not illuminated by sunlight.

Map of the Moon

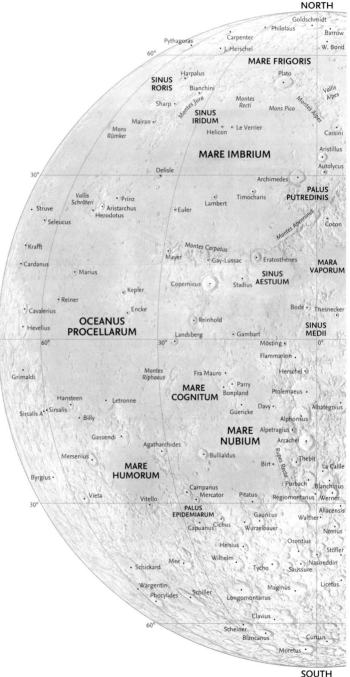

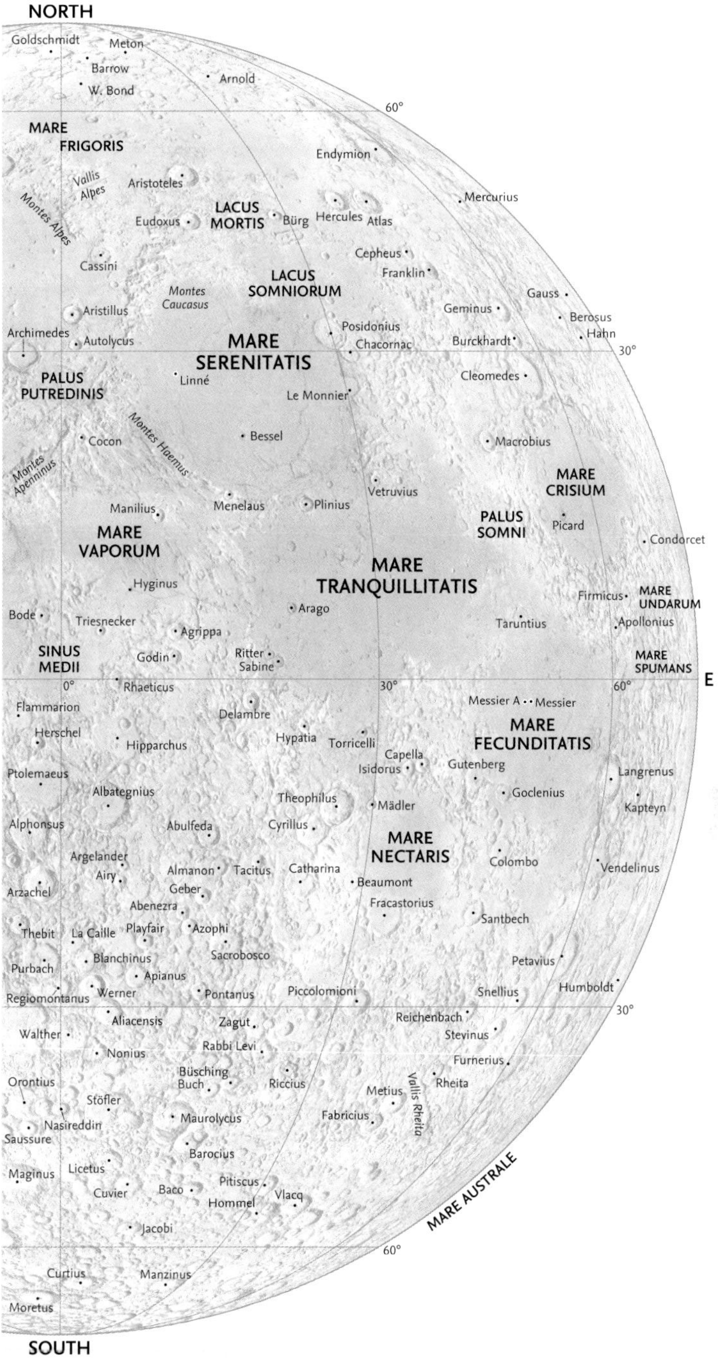

NORTH

Map labels (on the lunar map):

Goldschmidt · Meton
Barrow · Arnold
W. Bond
60°
MARE FRIGORIS
Endymion
Vallis Alpes · Aristoteles · Mercurius
Montes Alpes
Eudoxus · LACUS MORTIS · Bürg · Hercules · Atlas
Cassini · Cepheus
Montes Caucasus · LACUS SOMNIORUM · Franklin · Gauss
Aristillus · Posidonius · Geminus · Berosus
Archimedes · Autolycus · MARE SERENITATIS · Chacornac · Burckhardt · Hahn · 30°
PALUS PUTREDINIS · Linné · Cleomedes
Montes Haemus · Le Monnier
Cocon · Bessel · Macrobius
Montes Apenninus · Vetruvius · MARE CRISIUM
Manilius · Menelaus · Plinius · PALUS SOMNI · Picard
MARE VAPORUM · MARE TRANQUILLITATIS · Condorcet
Hyginus · Arago · Firmicus · MARE UNDARUM
Bode · Triesnecker · Agrippa · Taruntius · Apollonius
SINUS MEDII · Godin · Ritter · MARE SPUMANS
0° · Rhaeticus · Sabine · 30° · 60° · E
Flammarion · Delambre · Messier A · Messier
Herschel · Hipparchus · Hypatia · Torricelli · MARE FECUNDITATIS
Ptolemaeus · Capella · Gutenberg · Langrenus
Albategnius · Theophilus · Isidorus · Goclenius · Kapteyn
Alphonsus · Abulfeda · Cyrillus · Mädler
Argelander · MARE NECTARIS · Colombo
Airy · Almanon · Tacitus · Catharina · Beaumont · Vendelinus
Arzachel · Geber · Fracastorius
Abenezra · Santbech
Thebit · La Caille · Playfair · Azophi · Sacrobosco
Purbach · Blanchinus · Apianus · Petavius · Humboldt
Regiomontanus · Werner · Pontanus · Piccolomini · Snellius
Aliacensis · Zagut · Reichenbach · 30°
Walther · Nonius · Rabbi Levi · Stevinus · Furnerius
Orontius · Büsching · Buch · Riccius · Metius · Rheita
Stöfler · Fabricius · Vallis Rheita
Nasireddin · Maurolycus
Saussure · Barocius
Maginus · Licetus · Pitiscus · Vlacq
Cuvier · Baco · Hommel
Jacobi
60°
Curtius · Manzinus
Moretus

SOUTH

Macrobius	4:18
Mädler	5:19
Maginus	8:22
Manilius	7:21
Mare Crisium	2-3:16-17
Maurolycus	6:20
Mercator	10:24
Metius	4:18
Meton	6:20
Mons Pico	8:22
Mons Piton	8:22
Mons Rümker	12:26
Montes Alpes	6-8:21
Montes Apenninus	8
Orontius	8:22
Pallas	8:22
Petavius	3:17
Philolaus	9:23
Piccolomini	5:19
Pitatus	8:22
Pitiscus	5:19
Plato	8:22
Plinius	6:20
Posidonius	5:19
Proclus	14:18
Ptolemaeus	8:22
Purbach	8:22
Pythagoras	12:26
Rabbi Levi	6:20
Reinhold	9:23
Rima Ariadaeus	6:20
Rupes Recta	8
Saussure	8:22
Scheiner	10:24
Schickard	12:26
Sinus Iridum	10:24
Snellius	3:17
Stöfler	7:21
Taruntius	4:18
Thebit	8:22
Theophilus	5:19
Timocharis	8:22
Triesnecker	6-7:21
Tycho	8:22
Vallis Alpes	7:21
Vallis Schröteri	11:25
Vlacq	5:19
Walther	7:21
Wargentin	12:27
Werner	7:21
Wilhelm	9:23
Zagut	6:20

Introduction to the Month-by-Month Guide

The monthly charts

The pages devoted to each month contain a pair of charts showing the appearance of the night sky, looking north and looking south. The charts (as with all the charts in this book) are drawn for the latitude of 50°N, so observers farther north will see slightly more of the sky on the northern horizon, and slightly less on the southern. These areas are, of course, those most likely to be affected by poor observing conditions caused by haze, mist or smoke. In addition, stars close to the horizon are always dimmed by atmospheric absorption, so sometimes the faintest stars marked on the charts may not be visible.

The three times shown for each chart require a little explanation. The charts are drawn to show the appearance at 23:00 GMT for the 1st of each month. The same appearance will apply an hour earlier (22:00 GMT) on the 15th, and yet another hour earlier (21:00 GMT) at the end of the month (shown as the 1st of the following month). GMT is identical to the Universal Time (UT) used by astronomers around the world. In Europe, Summer Time is introduced in March, so the March charts apply to 23:00 GMT on March 1, 22:00 GMT on March 15, but 22:00 BST (British Summer Time) on April 1. The change back from Summer Time (in Europe) occurs in October, so the charts for that month apply to 00:00 BST for October 1, 23:00 BST for October 15, and 21:00 GMT for November 1.

The charts may be used for earlier or later times during the night. To observe two hours earlier, use the charts for the preceding month; for two hours later, the charts for the next month.

Meteors

Details of specific meteor showers are given in the months when they come to maximum, regardless of whether they begin or end in other months. Note that not all the respective radiants are marked on the charts for that

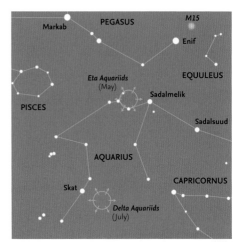

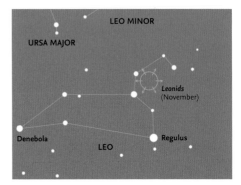

particular month, because the radiants may be below the horizon, or lie in constellations that are not readily visible during the month of maximum. For this reason, special charts for the Eta and Delta Aquariids (May and July, respectively) and the Leonids (November) are given here. As explained earlier, however, meteors from such showers may still be seen, because the most effective region for seeing meteors is some 40–45° away from the radiant, and that area of sky may well be above the horizon. A table of the best meteor showers visible during the year is also given here.

Meteors that are brighter than magnitude -4 (approximately the maximum magnitude reached by Venus) are known as **fireballs** or

Shower	Dates of activity 2018	Date of maximum 2018	Possible hourly rate
Quadrantids	January 1–10	January 3–4	120
April Lyrids	April 16–25	April 22–23	18
Eta Aquariids	April 19 to May 26	May 6–7	55
Alpha Capricornids	July 11 to August 10	July 27–28	5
Perseids	July 13 to August 26	August 11–12	100
Delta Aquariids	July 21 to August 23	July 28–29	< 20
Alpha Aurigids	August to October	August 28 & September 15	10
Southern Taurids	September 7 to November 19	October 9–10	< 5
Northern Taurids	October 19 to December 10	November 11–12	< 5
Orionids	October 4 to November 14	October 21–22	25
Leonids	November 5–30	November 17–18	< 15
Geminids	December 4–16	December 13–14	100+
Ursids	December 17–23	December 21–22	< 10

bolides. An example is shown on page 18. Fireballs sometimes cause sonic booms that may be heard some time after the meteor is seen.

The photographs

As an aid to identification – especially as some people find it difficult to relate charts to the actual stars they see in the sky – one or more photographs of constellations visible in certain specific months are included. It should be noted, however, that because of the limitations of the photographic and printing processes, and the differences between the sensitivity of different individuals to faint starlight (especially in their ability to detect different colours), and the degree to which they have become adapted to the dark, the apparent brightness of stars in the photographs will not necessarily precisely match that seen by any one observer.

The Moon calendar

The Moon calendar is largely self-explanatory. It shows the phase of the Moon for every day of the month, with the exact times (in Universal Time) of New Moon, First Quarter, Full Moon and Last Quarter. Because the times are calculated from the Moon's actual orbital parameters, some of the times shown will, naturally, fall during daylight, but any difference is too small to affect the appearance of the Moon on that date. Also shown is the *age* of the Moon (the day in the *lunation*),

beginning at New Moon, which may be used to determine the best time for observation of specific lunar features.

The Moon

The section on the Moon includes details of any lunar or solar eclipses that may occur during the month (visible from anywhere on Earth). Similar information is given about any important occultations. Mainly, however, this section summarizes when the Moon passes close to planets or the five prominent stars close to the ecliptic. The dates when the Moon is closest to the Earth (at *perigee*) and farthest from it (at *apogee*) are shown in the monthly calendars, and only mentioned here when they are particularly significant, such as the nearest and farthest during the year.

The planets and minor planets

Brief details are given of the location, movement and brightness of the planets from Mercury to Saturn throughout the month. None of the planets can, of course, be seen when they are close to the Sun, so such periods are generally noted. All of the planets may sometimes lie on the opposite side of the Sun to the Earth (at superior conjunction), but in the case of the inferior planets, Mercury and Venus, they may also pass between the Earth and the Sun (at inferior conjunction) and are invisible for a period of time, the length of which varies from conjunction to conjunction.

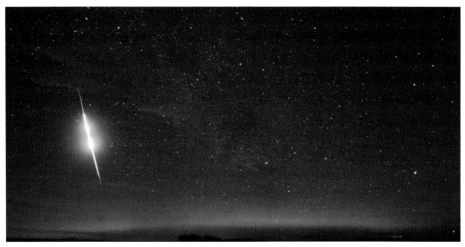

A fireball (with flares approximately as bright as the Full Moon), photographed against a weak auroral display by D. Buczynski from Portmahomack in Scotland on 22 January 22, 2017.

Those two planets are normally easiest to see around either eastern or western elongation, in the evening or morning sky, respectively. Not every elongation is favourable, so although every elongation is listed, only those where observing conditions are favourable are shown in the individual diagrams of events.

The dates at which the superior planets reverse their motion (from direct motion to **retrograde**, and retrograde to direct) and of opposition (when a planet generally reaches its maximum brightness) are given. Some planets, especially distant Saturn, may spend most or all of the year in a single constellation. Jupiter and Saturn are normally easiest to see around opposition, which occurs every year. Mars, by contrast, moves relatively rapidly against the background stars and in some years never comes to opposition.

Uranus is not always included in the monthly details because it is generally at the limit of naked-eye visibility (magnitude 5.7–5.9), although bright enough to be visible in binoculars, or even with the naked eye under exceptionally dark skies. Its path in 2018 is shown on the special chart in October. It comes to opposition on October 24 in the constellation of Aries. Full Moon occurs that day, so the planet, at magnitude 5.7, will be difficult (if not impossible) to detect. It is at the same magnitude for an extended period of the year (from August to December) and should be visible when free from interference by moonlight.

Similar considerations apply to Neptune, although this is always fainter (magnitude 7.8–8.0 in 2018), but still visible in most binoculars. It reaches opposition at mag. 7.8 on September 7 in Aquarius. New Moon occurs two days later, so Neptune should be detectable, especially a few days after that date. Again, Neptune's path in 2018 and its position at opposition are shown in a chart in September.

Charts for the three brightest minor planets that come to opposition in 2018 are shown in the relevant month: Ceres (mag. 6.9) on January 31, Vesta (mag. 5.3) on June 19, and Juno (mag. 7.4) on November 17.

The ecliptic charts

Although the ecliptic charts are primarily designed to show the positions and motions of the major planets, they also show the motion of the Sun during the month. The light-tinted area shows the area of the sky that is invisible during daylight, but the darker area gives an indication of which constellations are likely to be visible at some time of the night.

The closer a planet is to the border between dark and light, the more difficult it will be to see in the twilight.

The monthly calendar

For each month, a calendar shows details of significant events, including when planets are close to one another in the sky, close to the Moon, or close to any one of five bright stars that are spaced along the ecliptic. The times shown are given in Universal Time (UT), always used by astronomers throughout the year, and which is identical to Greenwich Mean Time (GMT). So during the summer months, they do not show Summer Time, which will always be one hour later than the time shown.

The diagrams of interesting events

Each month, a number of diagrams show the appearance of the sky when certain events take place. However, the exact positions of celestial objects and their separations greatly depend on the observer's position on Earth. When the Moon is one of the objects involved, because it is relatively close to Earth, there may be very significant changes from one location to another. Close approaches between planets or between a planet and a star are less affected by changes of location, which may thus be ignored.

The diagrams showing the appearance of the sky are drawn for the latitude of London, so will be approximately correct for most of Britain and Europe. However, for an observer farther north (say Edinburgh), a planet or star listed as being north of the Moon will appear even farther north, whereas one south of the Moon will appear closer to it – or may even be hidden (occulted) by it. For an observer farther south than London, there will be corresponding changes in the opposite direction: for a star or planet south of the Moon the separation will increase, and for one north of the Moon the separation will decrease. This is particularly important when occultations occur, which may be visible from one location, but not another. However, it does not apply to any occultations visible from Britain in 2018.

Ideally, details should be calculated for each individual observer, but this is obviously impractical. In fact, positions and separations are actually calculated for a theoretical observer located at the centre of the Earth.

So the details given regarding the positions of the various bodies should be used as a guide to their location. A similar situation arises with the times that are shown. These are calculated according to certain technical criteria, which need not concern us here. However, they do not necessarily indicate the exact time when two bodies are closest together. Similarly, dates and times are given, even if they fall in daylight, when the objects are likely to be completely invisible. However, such times do give an indication that the objects concerned will be in the same general area of the sky during both the preceding, and the following nights.

Key to the symbols used on the monthy star maps.

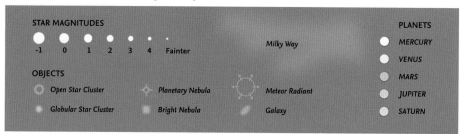

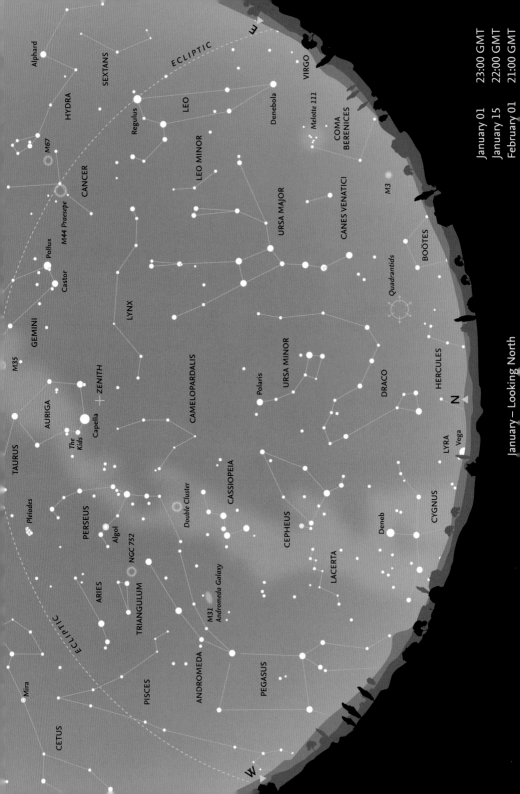

January – Looking North

January 01   23:00 GMT
January 15   22:00 GMT
February 01   21:00 GMT

January – Looking North

Most of the important circumpolar constellations are easy to see in the northern sky at this time of year. **Ursa Major** stands more-or-less vertically above the horizon in the northeast, with the zodiacal constellation of **Leo** rising in the east. To the north, the stars of **Ursa Minor** lie below **Polaris** (the Pole Star). The head of **Draco** is low on the northern horizon, but may be difficult to see unless observing conditions are good. Both **Cepheus** and **Cassiopeia** are readily visible in the northwest, and even the faint constellation of **Camelopardalis** is high enough in the sky for it to be easily visible.

Near the zenith is the constellation of **Auriga** (the Charioteer), with brilliant **Capella** (α Aurigae), directly overhead. Slightly to the west of Capella lies a small triangle of fainter stars, known as 'The Kids'. (Ancient mythological representations of Auriga show him carrying two young goats.) Together with the northernmost bright star in Taurus, β Tauri, the body of Auriga forms a large pentagon on the sky, with The Kids lying on the western side. Farther down towards the west are the constellations of **Perseus** and **Andromeda**, and the Great Square of **Pegasus** is approaching the horizon.

Meteors

One of the strongest and most consistent meteor showers of the year occurs in January: the **Quadrantids**, which are visible January 1–10, with maximum on January 3–4. They are brilliant, bluish and yellowish-white meteors and at maximum may even reach a rate of 120 meteors per hour. At maximum, the Moon is waning gibbous, just after Full Moon, so there will be considerable interference from moonlight. The parent object is minor planet 2003 EH$_1$.

The shower is named after the former constellation **Quadrans Muralis** (the Mural Quadrant), an early form of astronomical instrument.

The Quadrantid meteor radiant is now within the northernmost part of **Boötes**, roughly halfway between θ Boötis and τ Herculis.

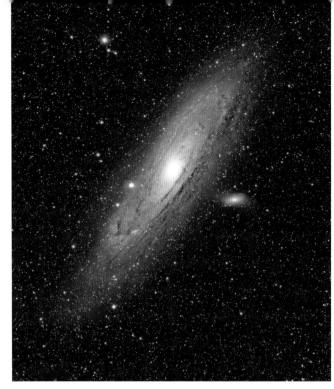

M31, the great Andromeda Galaxy, is about 220,000 light-years across, but has recently been found to be surrounded by an otherwise invisible halo of hot (10,000 to 100,000 degrees) ionized hydrogen at least five times that diameter, reaching halfway to our own Galaxy.

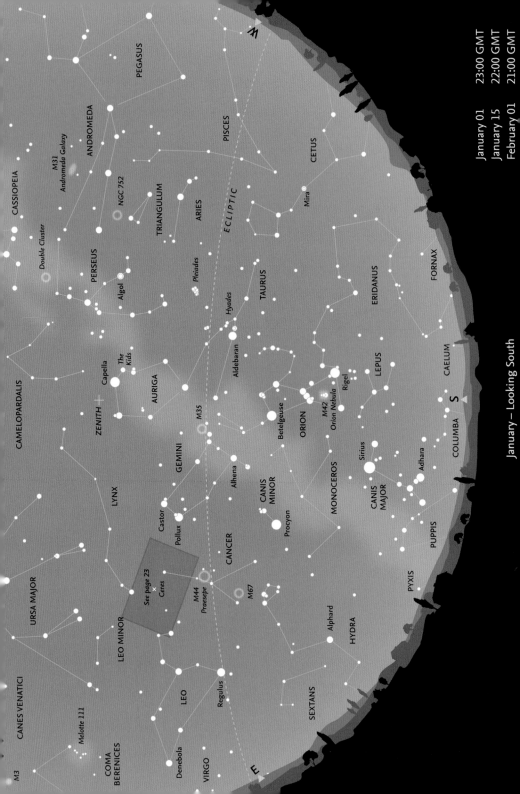

January 01   23:00 GMT
January 15   22:00 GMT
February 01   21:00 GMT

January – Looking South

January – Looking South

At this time of year the southern sky is dominated by **Orion**. This is the most prominent constellation during the winter months, when it is visible at some time during the night. (A photograph of Orion appears on page 87.) It has a highly distinctive shape, with a line of three stars that form the 'Belt'. To most observers, the bright star at the northeastern corner of the constellation, **Betelgeuse** (α Orionis), shows a reddish tinge, in contrast to the brilliant bluish-white colour of the bright star at the southwestern corner, **Rigel** (β Orionis). The three stars of the belt lie directly south of the celestial equator. A vertical line of three 'stars' forms the 'Sword' that hangs to the south of the Belt. With good viewing conditions, the central 'star' appears as a hazy spot, even to the naked eye. This is actually the Orion Nebula. Binoculars will reveal the four stars of the '**Trapezium**', which illuminate the nebula.

The line of Orion's Belt points up to the northwest towards **Taurus** (the Bull) and below orange-tinted **Aldebaran** (α Tauri). Close to Aldebaran, there is a conspicuous 'V' of stars, pointing down to the southwest, called the **Hyades** cluster. (Despite appearances, Aldebaran is not part of the cluster.) Farther along, the same line from Orion

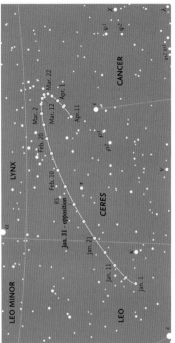

The path of minor planet (1) Ceres which is at opposition (mag. 6.9) on January 31. Background stars are shown down to magnitude 8.5.

passes below a bright cluster of stars, the **Pleiades**, or Seven Sisters. Even the smallest pair of binoculars reveals this cluster to be a beautiful group of bluish-white stars. The two most conspicuous of the other stars in Taurus lie directly above Orion, and form an elongated triangle with Aldebaran. The northernmost, β Tauri, was once considered to be part of the constellation of Auriga.

The Moon's phases for January

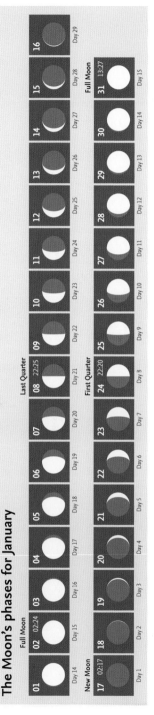

Full Moon		
01	02 02:24	03
Day 14	Day 15	Day 16
New Moon		
17 02:17	18	19
Day 1	Day 2	Day 3

January – Moon and Planets

The Earth

The Earth reaches perihelion (the closest point to the Sun in its annual orbit) on 3 January 2018, at 05:35 Universal Time. Its distance is then 0.9833 AU (147,097,192 km).

The Moon

In January, the Moon has both the closest perigee of the year, on January 1, at a distance of 356,565 km, and the most distant apogee, on January 15, at a distance of 406,464 km. On January 2, at Full, the Moon is in *Gemini*. By January 5 it is close to *Regulus*. (There is an occultation, visible from Arctic Canada.) On January 7, just before Last Quarter, it is near *Spica* in the early morning sky. The Moon (as waning crescent) is close to *Antares* on January 13 and (as waxing gibbous) near

Aldebaran on January 27. There is a total lunar eclipse, visible from Australia and eastern Asia, on January 31.

The planets

Mercury is at greatest western elongation (magnitude -0.3) on January 1 and rapidly moves towards the Sun. *Venus* is at superior conjunction (on the far side of the Sun) on January 9. *Mars* is in *Libra*, moving eastwards and increasing from mag. 1.5 to mag. 1.2 by January 31. *Jupiter* is also in *Libra*, moving more slowly eastwards, brightening slightly from mag. -1.8 to -2.0 over the month. The two planets are closest on January 7. *Saturn* is in *Sagittarius*, far too close to the Sun to be visible. *Uranus* (magnitude 5.8) is in *Pisces*, where it remains throughout the year. Similarly, *Neptune* remains in *Aquarius* for the whole year and is magnitude 7.9–8.0 in *January*. Minor planet (*1*) *Ceres* comes to opposition just inside *Cancer* (at magnitude 6.9) on January 31 at Full Moon. (A chart is shown on page 23.)

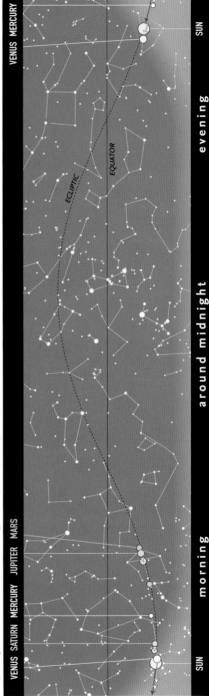

The path of the Sun and the planets along the ecliptic in January.

Calendar for January

01–10		Quadrantid meteor shower
01	19:58	Mercury greatest elongation (22.7°W, mag. –0.3)
01	21:49	Moon at perigee (closest of year, 356,565 km)
02	02:24	Full Moon
02	23:28	Pollux 8.6°N of Moon
03–04		Quadrantid shower maximum
03	05:35	Earth at perihelion (147,097,192 km = 0.983284 AU)
05	07:24	Regulus 0.9°S of Moon
07	00:39	Mars 0.21°S of Jupiter
08	22:25	Last Quarter
09	03:38	Spica 7.4°S of Moon
09	07:02	Venus superior conjunction
11	05:58	Jupiter 4.3°S of Moon
11	10:03	Mars 4.6°S of Moon
13	00:17	Antares 9.5°S of Moon
13	08:00*	Mercury 0.6°S of Saturn
15	02:10	Moon at apogee (greatest of year, 406,464 km)
15	01:57	Saturn 2.6°S of Moon
15	07:22	Mercury 3.7°S of Moon
17	02:17	New Moon
17	07:39	Venus 2.8°S of Moon
24	22:20	First Quarter
27	10:34	Aldebaran 0.7°S of Moon
30	09:57	Moon at perigee (359,000 km)
31	12:51	Ceres at opposition (mag. 6.9)
31	13:27	Full Moon
31	13:30	Total lunar eclipse (Pacific, Australia, E. Asia)

These objects are close together for an extended period around this time.

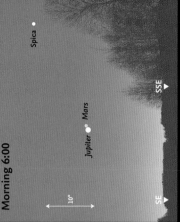

Morning 7:30

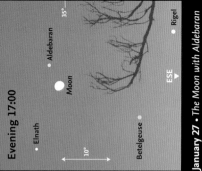

Morning 6:00

January 5 • The Moon between Regulus and γ Leonis (Algieba) in the west.

January 7 • Jupiter and Mars are close together in the early morning sky. Spica is nearby.

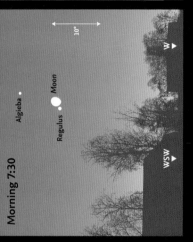

Morning 7:00

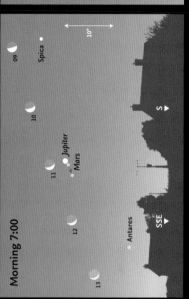

Evening 17:00

January 9–13 • The Moon passes Spica, Jupiter, Mars and Antares, shortly before sunrise.

January 27 • The Moon with Aldebaran high in the east-southeast.

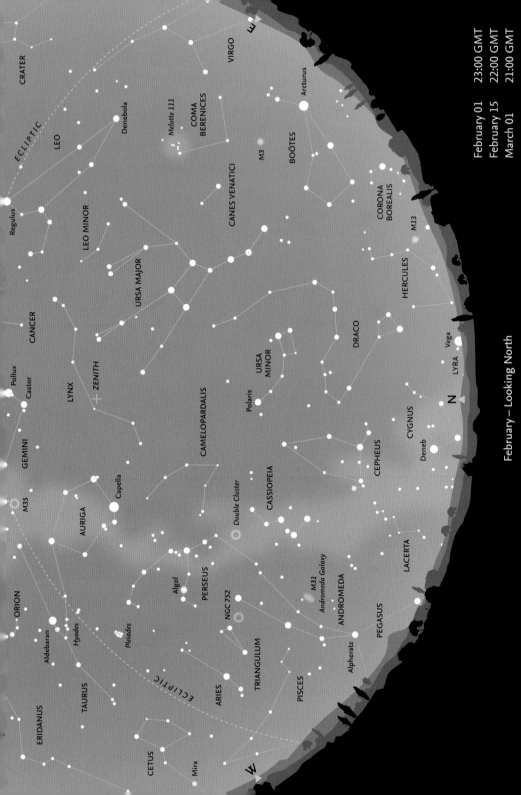

February 01   23:00 GMT
February 15   22:00 GMT
March 01   21:00 GMT

February – Looking North

February – Looking North

The months of January and February are probably the best time for seeing the section of the Milky Way that runs in the northern and western sky from **Cygnus**, low on the northern horizon, through **Cassiopeia**, **Perseus** and **Auriga** and then down through **Gemini** and **Orion**. Although not as readily visible as the denser star clouds of the summer Milky Way, on a clear night so many stars may be seen that even a distinctive constellation such as **Cassiopeia** is not immediately obvious.

The head of **Draco** is now higher in the sky and easier to recognize. **Deneb**, (α Cygni), the brightest star in **Cygnus**, may just be visible almost due north at midnight, early in the month, if the sky is very clear and the horizon clear of obstacles. **Vega** (α Lyrae) in **Lyra** is so low that it is difficult to see, but may become visible later in the night. The constellation of **Boötes** – sometimes described as shaped like a kite, an ice-cream cone, or the letter 'P' – with orange-tinted **Arcturus** (α Boötis), is beginning to clear the eastern horizon. Arcturus, at magnitude -0.05, is the brightest star in the northern hemisphere. The inconspicuous constellation of **Coma Berenices** is now well above the horizon in the east. The concentration of faint stars at the northwestern corner somewhat resembles a tiny, detached portion of the Milky Way. This is Melotte 111, an open star cluster (which is sometimes called the Coma Cluster, but must not be confused with the important Coma Cluster of galaxies, Abell 1656, mentioned on page 41).

On the other side of the sky, in the northwest, most of the constellation of **Andromeda** is still easily seen, although **Alpheratz** (α Andromedae), the star that forms the northeastern corner of the Great Square of Pegasus – even though it is actually part of Andromeda – is becoming close to the horizon and more difficult to detect. High overhead, at the zenith, try to make out the very faint constellation of

Lynx. It was introduced in 1687 by the famous astronomer Johannes Hevelius to fill the largely blank area between **Auriga**, **Gemini** and **Ursa Major**, and is reputed to be so named because one needed the eyes of a lynx to detect it.

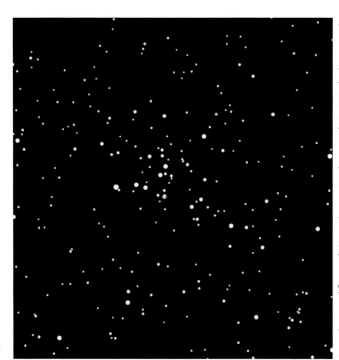

A very large, and frequently ignored, open star cluster, Melotte 111, also known as the Coma Cluster, is readily visible in the eastern sky during February.

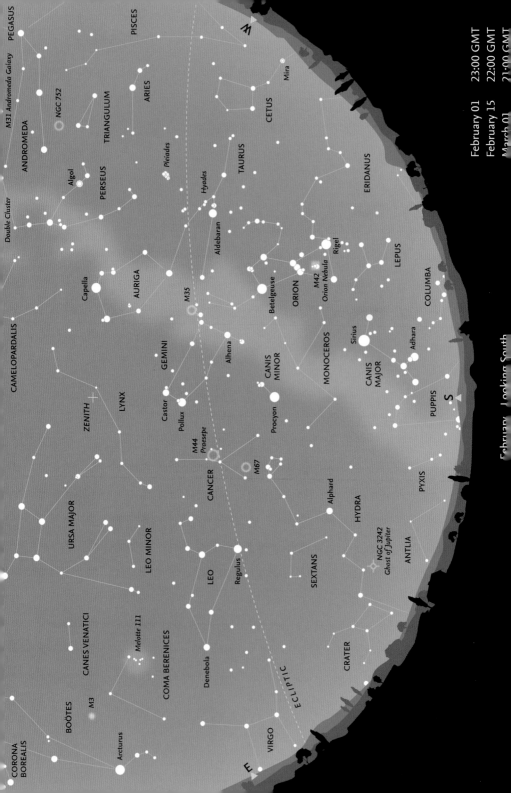

February 01   23:00 GMT
February 15   22:00 GMT
March 01   21:00 GMT

February, Looking South

February – Looking South

Apart from Orion, the most prominent constellation visible this month is **Gemini**, with its two lines of stars running southwest towards **Orion**. Many people have difficulty in remembering which is which of the two stars **Castor** and **Pollux**. Think of them in alphabetical order: Castor (α Geminorum), the fainter star, is closer to the North Celestial Pole. Pollux (β Geminorum) is the brighter of the two, but is farther away from the Pole. Castor is remarkable because it is actually a multiple system, consisting of no less than six individual stars.

Using Orion's belt as a guide, it points down to the southeast towards **Sirius**, the brightest star in the sky (at magnitude -1.4) in the constellation of **Canis Major**, the whole of which is now clear of the southern horizon. Forming an equilateral triangle with **Betelgeuse** in **Orion** and **Sirius** in Canis Major is **Procyon**, the brightest star in the small constellation of **Canis Minor**. Between Canis Major and Canis Minor is the faint constellation of **Monoceros**, which actually straddles the Milky Way, which, although faint, has many clusters in this area. Directly east of Procyon is the highly distinctive asterism of six stars that form the 'head' of **Hydra**, the largest

The constellation of Gemini. The two brightest stars are Castor and Pollux, visible on the left-hand side of the photograph.

of all 88 constellations, and which trails such a long way across the sky that it is only in mid-March around midnight that the whole constellation becomes visible.

The Moon's phases for February

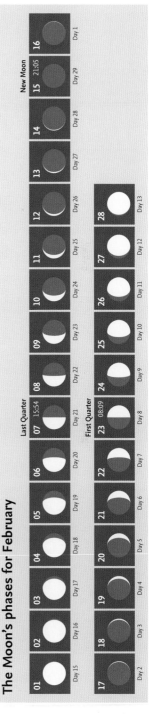

February – Moon and Planets

The Moon

On February 1, at 18:48 the Moon passes close to **Regulus** in **Leo**. On February 15 at New Moon, there is a partial solar eclipse, visible solely from Antarctica and the southern region of South America. That day it is in **Capricornus**, close to **Venus**. On February 23, the Moon is close to **Aldebaran** and part of an occultation is visible over a path from Central Europe to northern Asia.

Occultations

Of the bright stars near the ecliptic that may be occulted by the Moon (Aldebaran, Antares, Pollux, Regulus and Spica), **Aldebaran** is occulted nine times in 2018, and **Regulus** five times, none seen from Britain. (There are no occultations of Antares, Pollux or Spica in 2018.)

The planets

Mercury is too close to the Sun to be readily visible and is at superior conjunction, behind the Sun, on February 17. **Venus** is in the evening sky at magnitude -3.9, initially too close to the Sun to be visible, but may be glimpsed just before setting at the end of the month. **Mars** (magnitude 1.2 to 0.8) moves east in **Ophiuchus** in the early morning sky. **Jupiter** moves eastwards through **Libra**, brightening slightly from mag. -2.0 to -2.2. **Saturn** (mag. 0.6) is in **Sagittarius**, lost in the dawn sky. **Uranus** (mag. 5.9) is in **Pisces**, while **Neptune** (mag. 8.0) is in **Aquarius**, lost in daylight.

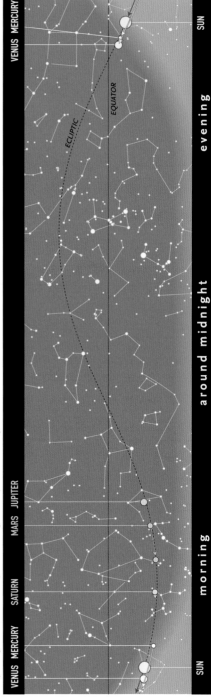

The path of the Sun and the planets along the ecliptic in February.

for February

Regulus 1.0°S of Moon

Spica 7.4°S of Moon

Last Quarter

Jupiter 4.3°S of Moon

Mars 4.4°S of Moon

Antares 9.5°S of Moon

Moon at apogee (405,700 km)

Saturn 2.7°S of Moon

Mars 5°N of Antares

Mercury 1.1°S of Moon

Partial solar eclipse
(Antarctica, Southern S. America)

New Moon

Venus 0.6°N of Moon

Mercury superior conjunction

First Quarter

Aldebaran 0.7°S of Moon

Pollux 8.6°N of Moon

Moon at perigee (363,900 km)

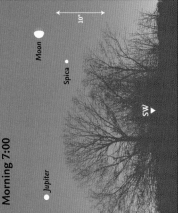

Evening 20:00

Algieba •

Moon

• Regulus

Alphard •

ESE

E

10°

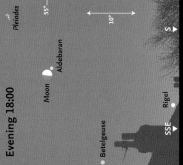

Morning 7:00

Jupiter

Moon

Spica •

SW

10°

February 1 • *The Moon and Regulus are close together when they rise in the east.*

February 5 • *The Moon with Spica in the southwest. Jupiter is farther to the south.*

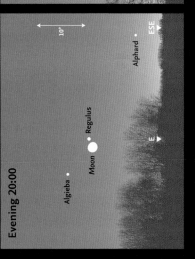

Morning 6:30

Sabik •

10

Saturn

11

• Saturn

09

Mars

• Antares

08

Jupiter

07

S

SSE

10°

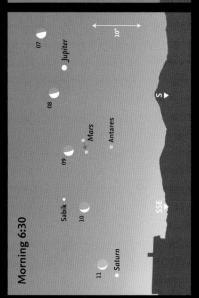

Evening 18:00

• Betelgeuse

Moon

• Aldebaran

Pleiades

Rigel

SSE

S

55°

10°

February 7–11 • *The Moon passes Jupiter, Mars, and Saturn, before sunrise. Mars and Antares are now almost equally bright.*

February 23 • *The Moon is close to Aldebaran. Try to spot the faint Pleiades.*

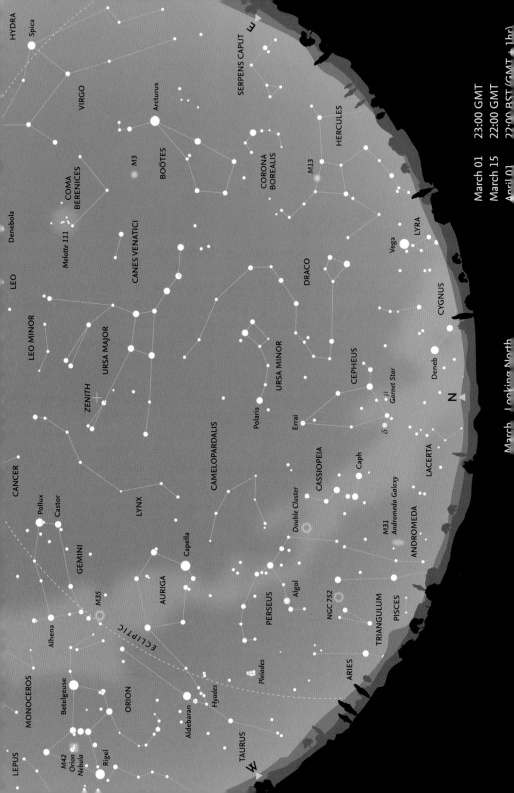

March – Looking North

March 01   23:00 GMT
March 15   22:00 GMT
April 01   22:00 BST (GMT + 1hr)

March – Looking North

In March, the Sun crosses the celestial equator on Tuesday, March 20, at the vernal equinox, when day and night are of almost exactly equal length, and the season of spring is considered to have begun. (The hours of daylight and darkness change most rapidly around the equinoxes, in March and September.) It is also in March that Summer Time begins in Europe (on Sunday, March 25) so the charts show the appearance at 23:00 GMT for March 1 and 22:00 BST for April 1. (In the USA, Daylight Saving Time is introduced two weeks earlier, on Sunday, March 11.)

Early in the month, the constellation of **Cepheus** lies almost due north, with the distinctive 'W' of **Cassiopeia** to its west. Cepheus lies across the border of the Milky Way and is often described as like the gable-end of a house or a church tower and steeple. Despite the large number of stars revealed at the base of the constellation by binoculars, one star stands out because of its deep red colour. This is Mu (μ) Cephei, also known as the **Garnet Star**, because of its striking colour. It is a truly gigantic star, a red supergiant, and one of the largest stars known. It is about 2400 times the diameter of the Sun, and if placed in the Solar System would extend beyond the orbit of Saturn. (Betelgeuse, in Orion, is also a red supergiant, but it is 'only' about 500 times the diameter of the Sun.)

Another famous, and very important star in Cepheus is **δ Cephei**, which is the prototype for the class of variable stars known as Cepheids. These giant stars show a regular variation in their luminosity, and there is a direct relationship between the period of the changes in magnitude and the stars' actual luminosity. From a knowledge of the period of any Cepheid, its actual luminosity – known as its absolute magnitude – may be derived. A comparison of its apparent magnitude on the sky and its absolute magnitude enables the star's exact distance to be

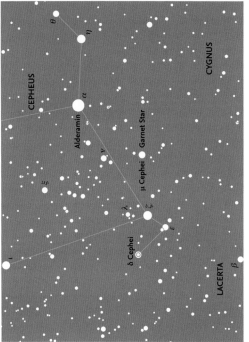

A finder chart for δ Cephei and μ Cephei (the Garnet Star). All stars brighter than magnitude 7.5 are shown.

determined. Once the distances to the first Cepheid variables had been established, examples in more distant galaxies provided information about the scale of the universe. Cepheid variables are the first 'rung' in the cosmic distance ladder. Both important stars are shown on the accompanying chart.

Below Cepheus to the east (to the right), it may be possible to catch a glimpse of **Deneb** (α Cygni), just above the horizon. Slightly farther round towards the northeast, **Vega** (α Lyrae) is marginally higher in the sky. From southern Britain, Deneb is just far enough north to be circumpolar (although difficult to see in January and February because it is so low on the northern horizon). Vega, by contrast, farther south, is completely hidden during the depths of winter.

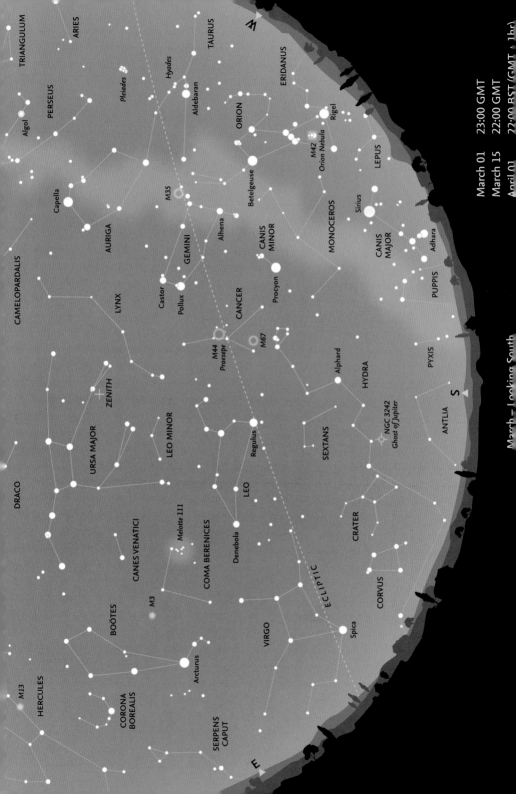

March – Looking South

March – Looking South

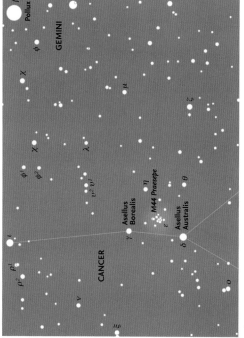

Due south at 22:00 at the beginning of the month, lying between the constellations of *Gemini* in the west and *Leo* in the east, and fairly high in the sky above the head of Hydra, is the faint, and rather undistinguished zodiacal constellation of *Cancer*. Rather like the triskelion, the symbol for the Isle of Man, it has three 'legs' radiating from the centre, where there is an open cluster, M44 or *Praesepe* ('the Manger' but also known as 'the Beehive'). On a clear night this cluster, known since antiquity, is just a hazy spot to the naked eye, but appears in binoculars as a group of dozens of individual stars.

Also prominent in March is the constellation of *Leo*, with the 'backward question mark' (or 'Sickle') of bright stars forming the head of the mythological lion. *Regulus* (α Leonis) – the 'dot' of the 'question mark' or the handle of the sickle and the brightest star in Leo – lies very close to the ecliptic and is one of the few first-magnitude stars that may be occulted by the Moon. Occultations occur on March 1 and March 28, but the first is visible from Arctic Canada only, and the second from Siberia.

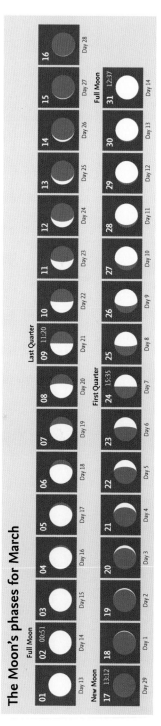

A finder chart for the open cluster M44 in Cancer. To the ancient Greeks and Romans, the two stars Asellus Borealis and Asellus Australis represented two donkeys feeding from Praesepe ('the Manger'). All stars brighter than magnitude 7.5 are shown.

The Moon's phases for March

01 Day 13	**02** 00:51 Full Moon Day 14	**03** Day 15	**04** Day 16	**05** Day 17	**06** Day 18
07 Day 19	**08** Day 20	**09** 11:20 Last Quarter Day 21	**10** Day 22	**11** Day 23	**12** Day 24
13 Day 25	**14** Day 26	**15** Day 27	**16** Day 28		
17 13:12 New Moon Day 29	**18** Day 1	**19** Day 2	**20** Day 3	**21** Day 4	**22** Day 5
23 Day 6	**24** 15:35 First Quarter Day 7	**25** Day 8	**26** Day 9	**27** Day 10	**28** Day 11
29 Day 12	**30** Day 13	**31** 12:37 Full Moon Day 14			

March – Moon and Planets

The Moon

On March 1, a day before Full Moon, there is an occultation of Regulus, visible from Arctic Canada. On March 4 the Moon is close to *Spica* in *Virgo* and during March 8 passes *Antares* in daylight. On March 21 and 22, it is in *Taurus*, near *Aldebaran*. (There is an occultation on March 22, which would be partially visible from the North Atlantic.) On March 31 (at Full Moon) it is again in *Virgo*, close to *Spica*.

The planets

Mercury, initially close to the Sun, reaches greatest eastern elongation (18.4°E, mag. -0.4) on March 15, but is low in the evening sky. *Venus* also moves rapidly eastwards of the Sun and reaches mag. -3.9 by March 31. *Mars*, initially in *Ophiuchus*, at magnitude 1.3, moves eastwards into *Sagittarius*, brightening from mag. 0.8 to 0.3 at the end of the month. *Jupiter* (mag. -2.2 to -2.4) is moving slowly eastwards in *Libra*, visible in the morning sky. *Saturn* is in *Sagittarius*, and might be glimpsed, low on the morning horizon. *Uranus* remains magnitude 5.9 in *Pisces* and *Neptune* magnitude 8.0 in *Aquarius*, too close to the Sun to be visible.

The path of the Sun and the planets along the ecliptic in March.

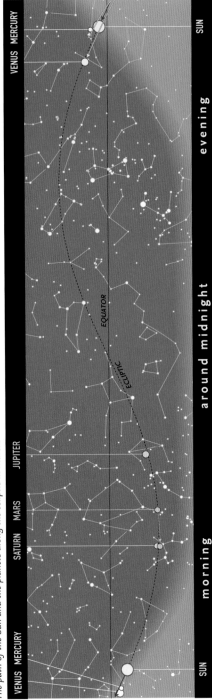

Calendar for March

01	05:33	Regulus 1.0°S of Moon
02	00:51	Full Moon
04	21:46	Spica 7.4°S of Moon
05	18:00*	Mercury 1.4°N of Venus
07	06:56	Jupiter 4.1°S of Moon
08	14:54	Antares 9.4°S of Moon
09	11:20	Last Quarter
10	00:38	Mars 3.8°S of Moon
11		Daylight Saving Time begins (North America)
11	02:21	Saturn 2.2°S of Moon
11	09:14	Moon at apogee (404,700 km)
15	15:10	Mercury greatest elongation (18.4°E, mag. -0.4)
17	13:12	New Moon
18	01:00*	Mercury 3.9°N of Venus
18	18:09	Mercury 7.7°N of Moon
18	19:05	Venus 3.8°N of Moon
20	16:15	Spring (vernal) eqinox
22	22:59	Aldebaran 0.9°S of Moon
24	15:35	First Quarter
25		Summer Time begins (Europe)
26	03:15	Pollux 8.5°N of Moon
26	17:17	Moon at perigee (369,100 km)
28	14:02	Regulus 1.0°S of Moon
31	12:37	Full Moon

*These objects are close together for an extended period around this time.

Morning 5:00

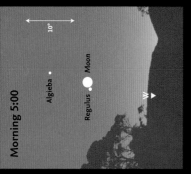

March 1 • The Moon with Regulus in the early morning, before they set.

Morning 6:00

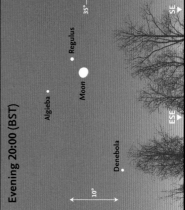
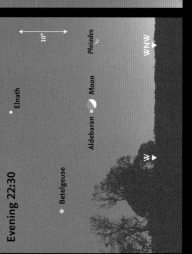

March 7-11 • The Moon passes Jupiter, Mars and Saturn. Mars is now becoming brighter than Antares.

Evening 22:30

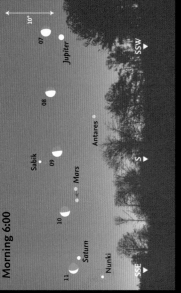

March 22 • The Moon and Aldebaran close together above the western horizon.

Evening 20:00 (BST)

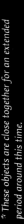

March 28 • The Moon in the company of Regulus and Algieba. Denebola is also nearby.

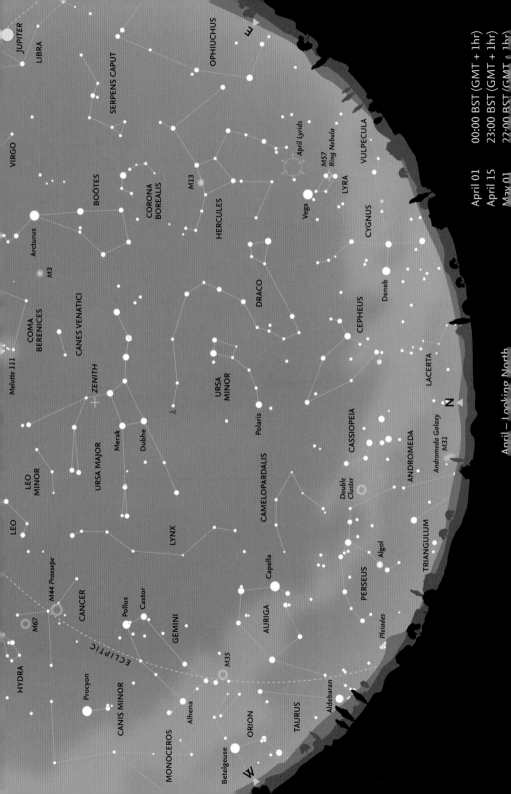

April – Looking North

April 01	00:00 BST (GMT + 1hr)
April 15	23:00 BST (GMT + 1hr)
May 01	22:00 BST (GMT + 1hr)

April – Looking North

Cygnus and the brighter regions of the Milky Way are now becoming visible, running more-or-less parallel with the horizon in the early part of the night. Rising in the northeast is the small constellation of **Lyra** and the distinctive 'Keystone' of **Hercules** above it. This asterism is very useful for locating the bright globular cluster M13 (see map on page 53), which lies on one side of the quadrilateral. The winding constellation of **Draco** weaves its way from the quadrilateral of stars that marks its 'head', on the border with Hercules, to end at λ Draconis between **Polaris** (α Ursae Minoris) and the 'Pointers', **Dubhe** and **Merak** (α and β Ursae Majoris, respectively). **Ursa Major** is 'upside down' high overhead, near the zenith. The constellation of **Gemini** stands almost vertically in the west. **Auriga** is still clearly seen in the northwest, but, by the end of the month, the southern portion of **Perseus** is starting to dip below the northern horizon. The very faint constellation of **Camelopardalis** lies in the northwest between Polaris and the constellations of **Auriga** and **Perseus**.

Meteors

A moderate meteor shower, the **Lyrids**, peaks on April 22–23. Although the hourly rate is not very high (about 18 meteors per hour), the meteors are fast and some leave persistent trains. This year the maximum occurs when the Moon is at and just after First Quarter, so conditions are reasonably favourable for observation. The parent object is the non-periodic comet C/1861 G1 (Thatcher). Another, stronger shower, the **Eta Aquariids**, begins to be active around April 19, and comes to maximum in May.

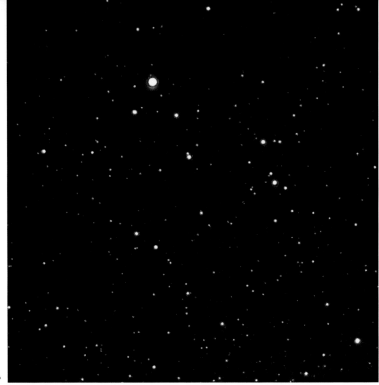

The constellation of Lyra is marked by the bright star Vega, which for most of Europe only dips below the northern horizon in mid-winter, with a distinctive quadrilateral of stars to its southeast.

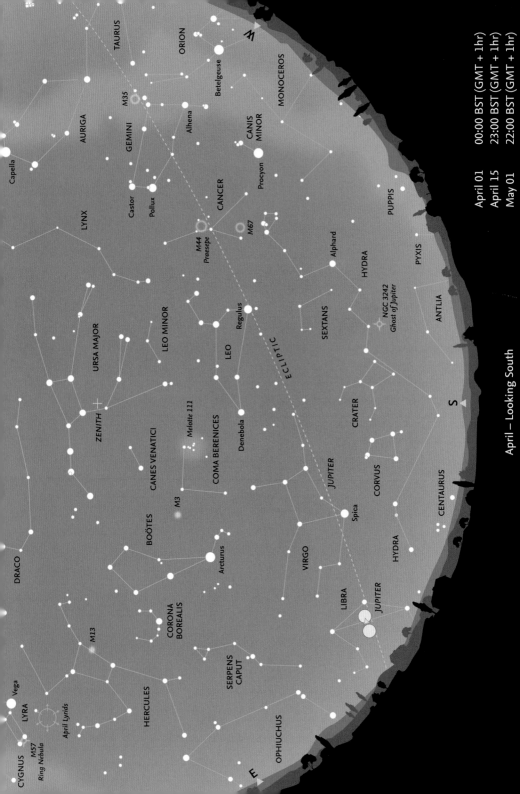

April – Looking South

April 01   00:00 BST (GMT + 1hr)
April 15   23:00 BST (GMT + 1hr)
May 01   22:00 BST (GMT + 1hr)

April – Looking South

Leo is the most prominent constellation in the southern sky in April, and vaguely looks like the creature after which it is named. **Gemini**, with **Castor** and **Pollux**, remains clearly visible in the west, and **Cancer** lies between the two constellations. To the east of Leo, the whole of **Virgo**, with **Spica** (α Virginis) its brightest star, is well clear of the horizon. Below Leo and Virgo, the complete length of **Hydra** is visible, running beneath both constellations, with **Alphard** (α Hydrae) halfway between Regulus and the southwestern horizon. Farther east, the two small constellations of **Crater** and the rather brighter **Corvus** lie between Hydra and Virgo.

Boötes and **Arcturus** are prominent in the eastern sky, together with the circlet of **Corona Borealis**, framed by Boötes and the neighbouring constellation of **Hercules**. Between Leo and Boötes lies the constellation of **Coma Berenices**, notable for being the location of the open cluster Melotte 111 (see page 27) and the Coma Cluster of galaxies (Abell 1656). There are about 1000 galaxies in this cluster, which is located near the North Galactic Pole, where we are looking out of the plane of the Galaxy and are thus able to see deep into space. Only about ten of the brightest galaxies in the Coma Cluster are visible with the largest amateur telescopes.

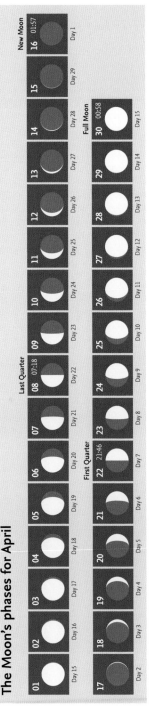

The distinctive constellation of Leo, with Regulus and 'The Sickle' on the west. Algieba (γ Leonis), north of Regulus, appearing double, is a multiple system of four stars.

The Moon's phases for April

01	02	03	04	05	06	07	08 07:18
Day 15	Day 16	Day 17	Day 18	Day 19	Day 20	Day 21	Day 22

First Quarter (22 21:46) · Last Quarter (08 07:18)

17	18	19	20	21	22 21:46	23	24
Day 2	Day 3	Day 4	Day 5	Day 6	Day 7	Day 8	Day 9

09	10	11	12	13	14	15	16 01:57
Day 23	Day 24	Day 25	Day 26	Day 27	Day 28	Day 29	Day 1

New Moon (16 01:57) · Full Moon (30 00:58)

25	26	27	28	29	30 00:58		
Day 10	Day 11	Day 12	Day 13	Day 14	Day 15		

April – Moon and Planets

The Moon

On April 1, the waning gibbous Moon (one day past Full) is close to *Spica* in *Virgo*. It is closest as the constellation begins to set in the early morning. In the early morning of April 4 it is close to *Jupiter* in *Libra* and by late evening is north of *Antares* in *Scorpius*. On April 21 and 22 (First Quarter) the Moon is in *Gemini* not far from *Castor* and *Pollux*. On April 24 it is close to *Regulus* in *Leo*. (There is an occultation, partially visible from northern and central Asia.) On April 28 (two days before Full Moon) it is again close to *Spica*, in the eastern sky.

The planets

Mercury passes inferior conjunction (between Earth and Sun) on April 1. It then reaches greatest western elongation (27.0°W, mag. 0.4) on April 29. *Venus* is bright (mag. -3.9) in April and rapidly moves from *Aries* into *Taurus* in the western sky. *Mars* moves eastwards across the constellation of *Sagittarius* over the month. On April 2, it is close to Saturn, low in the southern sky. *Jupiter* (mag. -2.4 to -2.5) starts retrograde motion early in April, and is in *Libra*. *Saturn* (mag. 0.4–0.5) is inside *Sagittarius*. *Uranus* (mag. 5.9) and *Neptune* (mag. 7.9) remain in *Pisces* and *Aquarius*, respectively, both too close to the Sun to be readily visible.

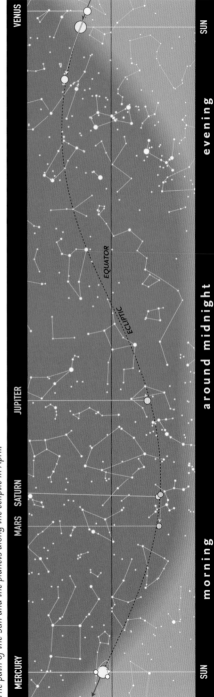

The path of the Sun and the planets along the ecliptic in April.

Calendar for April

01	07:29	Spica 7.3°S of Moon
01	17:53	Mercury inferior conjunction
02	12:00*	Saturn 1.3°N of Mars
03	14:14	Jupiter 3.9°S of Moon
04	23:35	Antares 9.3°S of Moon
07	12:50	Saturn 1.9°S of Moon
07	18:15	Mars 3.1°S of Moon
08	05:31	Moon at apogee (404,100 km)
08	07:18	Last Quarter
16–25		Lyrid meteor shower
16	01:57	New Moon
17	19:28	Venus 5.4°N of Moon
Apr.19–May26		Eta Aquarid meteor shower
19	05:10	Aldebaran 1.1°S of Moon
20	14:41	Moon at perigee (368,700 km)
22	08:32	Pollux 8.2°N of Moon
22	21:46	First Quarter
22–23		April Lyrid shower maximum
24	16:47	Venus 3.4°S of Moon
24	20:04	Regulus 1.2°S of Moon
28	15:31	Spica 7.3°S of Moon
29	18:24	Mercury greatest elongation (27.0°W, mag. 0.4)
30	00:58	Full Moon
30	17:16	Jupiter 3.8°S of Moon

* These objects are close together for an extended period around this time.

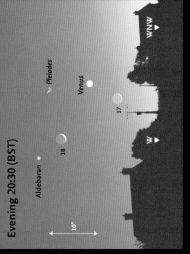

Morning 6:00 (BST)

April 2 • *Mars and Saturn above the southern horizon, 30 minutes before sunrise.*

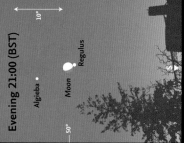

Evening 20:30 (BST)

April 17–18 • *The narrow crescent Moon is close to Venus on April 17. One day later it is approaching Aldebaran.*

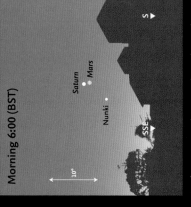

Evening 21:00 (BST)

April 21–22 • *The Moon with Castor and Pollux, high in the southwest.*

Evening 21:00 (BST)

April 24 • *Two days later the Moon is close to Regulus.*

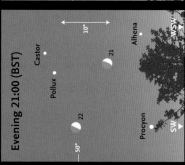

Evening 23:30 (BST)

April 30 • *The Moon with Jupiter, when they rise in the southeast.*

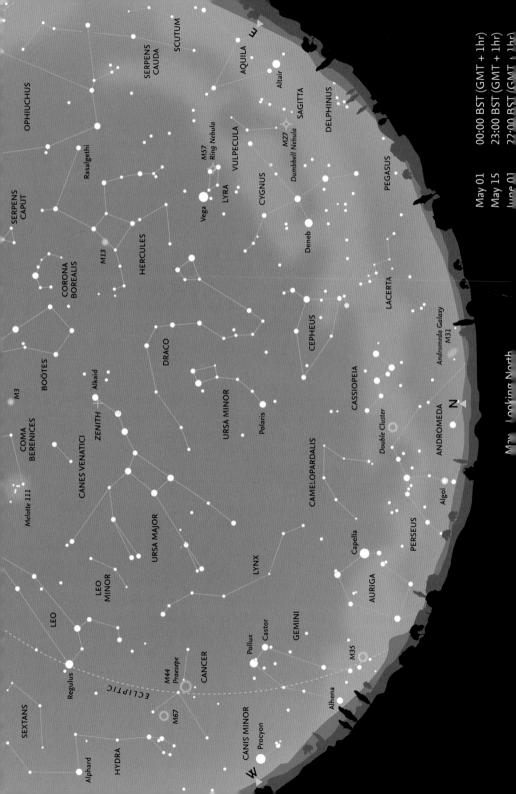

May, Looking North

May 01	00:00 BST (GMT + 1hr)
May 15	23:00 BST (GMT + 1hr)
June 01	22:00 BST (GMT + 1hr)

May – Looking North

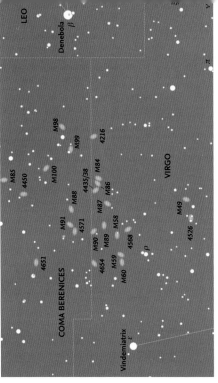

A finder chart for some of the brightest galaxies in the Virgo Cluster (see page 47). All stars brighter than magnitude 8.5 are shown.

Cassiopeia is now low over the northern horizon and, to its west, the southern portions of both **Perseus** and **Auriga** are becoming difficult to observe. **Gemini**, with **Castor** and **Pollux**, is sinking towards the western horizon.

In the east, two of the stars of the 'Summer Triangle', **Vega** in **Lyra** and **Deneb** in **Cygnus**, are clearly visible, and the third, **Altair** in **Aquila**, is beginning to climb above the horizon. The sprawling constellation of **Hercules** is high in the east and the brightest globular cluster in the northern hemisphere, M13, is visible to the naked eye on the western side of the 'Keystone'.

Later in the night (and in the month) the westernmost stars of **Pegasus** begin to come into view, while the stars of **Andromeda** are skimming the northeastern horizon. High overhead, **Alkaid** (η Ursae Majoris), the last star in the 'tail' of the Great Bear, is close to the zenith, while the main body of the constellation has swung round into the western sky.

Meteors

The **Eta Aquariids** are one of the two meteor showers associated with Comet 1P/Halley (the other being the **Orionids**, in October). The Eta Aquariids are not particularly favourably placed for northern-hemisphere observers, because the radiant is near the celestial equator, near the 'Water Jar' in **Aquarius**, well below the horizon until late in the night (around dawn). However, meteors may still be seen in the eastern sky even when the radiant is below the horizon. There is a radiant map for the Eta Aquariids on page 16.

Their maximum in 2018, on May 6–7, occurs when the Moon is waning gibbous, just before Last Quarter, so conditions are not ideal for observation but should be acceptable. Maximum hourly rate is about 55 per hour and a large proportion (about 25 per cent) of the meteors leave persistent trains.

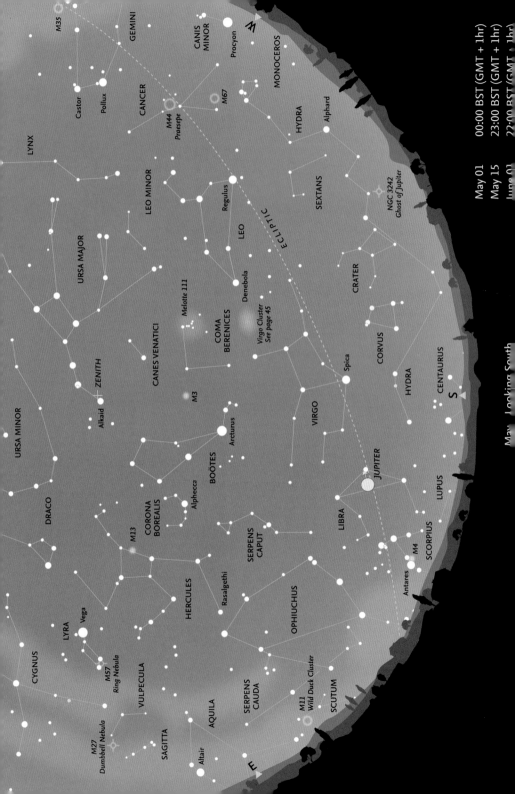

May, Looking South

May 01   00:00 BST (GMT + 1hr)
May 15   23:00 BST (GMT + 1hr)
June 01   22:00 BST (GMT + 1hr)

May – Looking South

Early in the night, the constellation of **Virgo**, with **Spica** (α Virginis), lies due south, with **Leo** and both **Regulus** and **Denebola** (α and β Leonis, respectively) to its west still well clear of the horizon. Later in the night, the rather faint zodiacal constellation of **Libra** becomes visible and, to its east, the ruddy star **Antares** (α Scorpii) begins to climb up over the horizon.

Virgo contains the nearest large cluster of galaxies, which is the centre of the Local Supercluster, of which the Milky Way galaxy forms part. The Virgo Cluster contains some 2000 galaxies, the brightest of which are visible in amateur telescopes.

Arcturus in **Boötes** is high in the south, with the distinctive circlet of **Corona Borealis** clearly visible to its east. The brightest star (α Coronae Borealis) is known as **Alphecca**. The large constellation of **Ophiuchus** (which actually crosses the ecliptic, and is thus the 'thirteenth' zodiacal constellation) is climbing into the eastern sky. Before the constellation boundaries were formally adopted by the International Astronomical Union in 1930, the southern region of Ophiuchus was

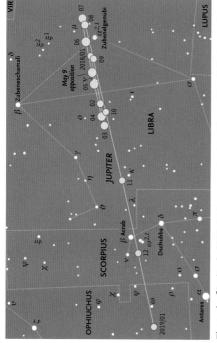

The path of Jupiter in 2018. Jupiter comes to opposition on May 9. Background stars are shown down to magnitude 6.5.

regarded as forming part of the constellation of Scorpius, which had been part of the Zodiac since antiquity.

The Moon's phases for May

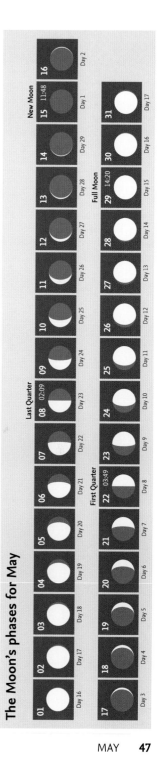

May – Moon and Planets

The Moon

On May 1, the waning gibbous Moon is northeast of **Jupiter** in **Libra**. It is close to **Saturn** on May 5 and **Mars** on May 6 (both in **Sagittarius**). It is very close to **Aldebaran** (α Tauri) on May 16 and **Regulus** in **Leo** on May 22, but there are no occultations. On May 29 (at Full Moon) it is in **Scorpius**, not far from **Antares**, as it was at the beginning of the month (on May 1). On May 31 it is in **Ophiuchus**.

The planets

Mercury, initially in **Pisces**, moves rapidly towards the Sun. **Venus** (mag. -3.9) moves rapidly eastwards from **Taurus** into **Gemini** in the course of the month, becoming a prominent object in the early evening sky. **Mars** is in **Sagittarius**, moving eastwards and entering **Capricornus** in mid-month, brightening from mag. -0.4 to -1.2, but visible only low in the east shortly after midnight. **Jupiter** (at magnitude -2.5), continues to retrograde in **Libra**, coming to opposition on May 9, as shown on the chart on page 47. **Saturn** (magnitude 0.4 to 0.2) is retrograding very slowly in **Sagittarius**. **Uranus**, now in **Aries**, continues direct motion at magnitude 5.9. **Neptune** is also moving slowly eastwards, at magnitude 7.9. Both planets are lost in the dawn sky.

The path of the Sun and the planets along the ecliptic in May.

Calendar for May

02	07:59	Antares 9.1°S of Moon
03	17:00*	Aldebaran 6.5°S of Venus
04	20:31	Saturn 1.7°S of Moon
06	00:35	Moon at apogee (404,500 km)
06	07:24	Mars 2.7°S of Moon
06–07		Eta Aquariid shower maximum
08	02:09	Last Quarter
09	00:39	Jupiter at opposition (mag. -2.5)
15	11:48	New Moon
16	13:28	Aldebaran 1.2°S of Moon
17	18:09	Venus 4.8°N of Moon
17	21:05	Moon at perigee (363,800 km)
19	14:41	Pollux 8.0°N of Moon
22	01:18	Regulus 1.5°S of Moon
22	03:49	First Quarter
25	21:39	Spica 7.4°S of Moon
27	17:40	Jupiter 4.0°S of Moon
29	14:20	Full Moon
29	15:16	Antares 9.0°S of Moon

* These objects are close together for an extended period around this time.

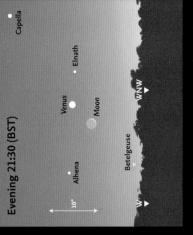

Morning 4:30 (BST)

May 5–6 • The Moon passes Saturn (mag. 0.2) and Mars (mag. -0.8).

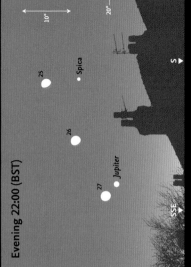

Evening 21:30 (BST)

May 17 • The narrow crescent Moon is close to Venus, which is now almost magnitude -4.

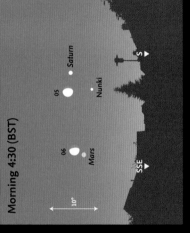

After midnight 1:00 (BST)

May 22 • The Moon with Regulus and Algieba, close to the western horizon.

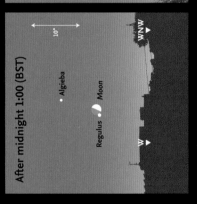

Evening 22:00 (BST)

May 25–27 • The Moon passes Spica and Jupiter.

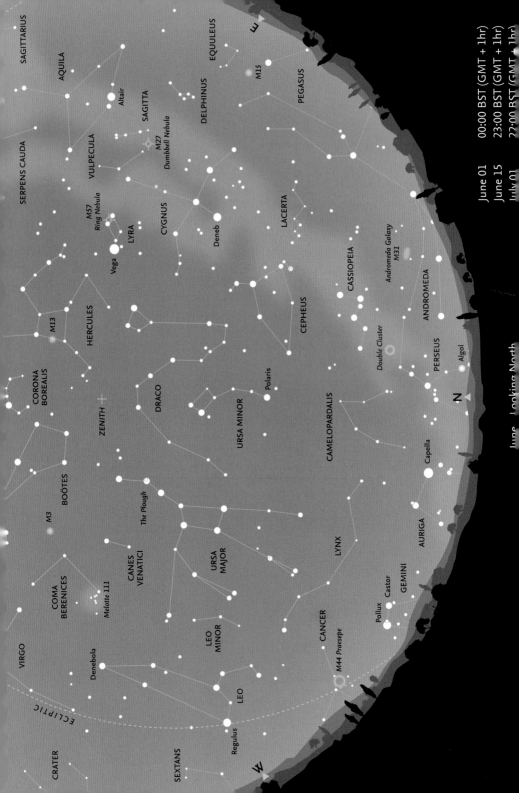

SAGITTARIUS
AQUILA
Altair
SAGITTA
EQUULEUS
M15
PEGASUS
DELPHINUS
SERPENS CAUDA
VULPECULA
M27
Dumbbell Nebula
LACERTA
M57
Ring Nebula
LYRA
CYGNUS
Vega
Deneb
ANDROMEDA
Andromeda Galaxy
M31
CASSIOPEIA
CEPHEUS
HERCULES
M13
Double Cluster
CORONA
BOREALIS
PERSEUS
Algol
DRACO
ZENITH
URSA MINOR
Polaris
N
M3
BOÖTES
CAMELOPARDALIS
Capella
The Plough
CANES
VENATICI
URSA
MAJOR
LYNX
AURIGA
COMA
BERENICES
Melotte 111
GEMINI
Castor
Pollux
VIRGO
Denebola
LEO
MINOR
CANCER
M44 *Praespe*
LEO
ECLIPTIC
Regulus
W
SEXTANS
CRATER

June – Looking North

Around summer solstice (June 21) even in southern England and Ireland a form of twilight persists throughout the night. Farther north, in Scotland, the sky remains so light that most of the fainter stars and constellations are invisible. There, even brighter stars, such as the seven stars making up the well-known asterism known as the *Plough* in *Ursa Major* may be difficult to detect except around local midnight, 00:00 UT (01:00 BST).

But there is one compensation during these light nights: even southern observers may be lucky enough to witness a display of noctilucent clouds (NLC). These are highly distinctive clouds shining with an electric-blue tint, observed in the sky in the direction of the North Pole. They are the highest clouds in the atmosphere, occurring at altitudes of 80–85 km, far above all other clouds. They are only visible during summer nights, for about a month or six weeks on either side of the solstice, when observers are in darkness, but the clouds themselves remain illuminated by sunlight, reaching them from the Sun, itself hidden below the northern horizon.

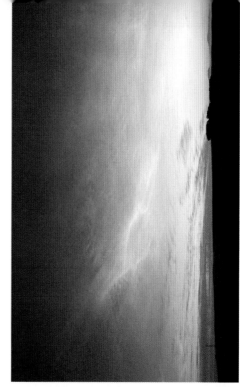

Noctilucent clouds, photographed on the night of 5–6 July 2016, from Portmahomack, Ross-shire, Scotland, by Denis Buczynski.

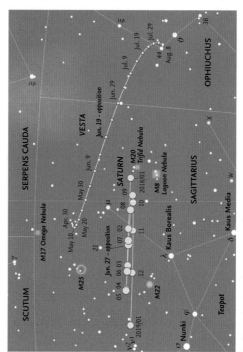

A chart showing the positions of Saturn and minor planet (4) Vesta around the dates of opposition (June 27 and June 19, respectively). All stars brighter than magnitude 7.0 are shown.

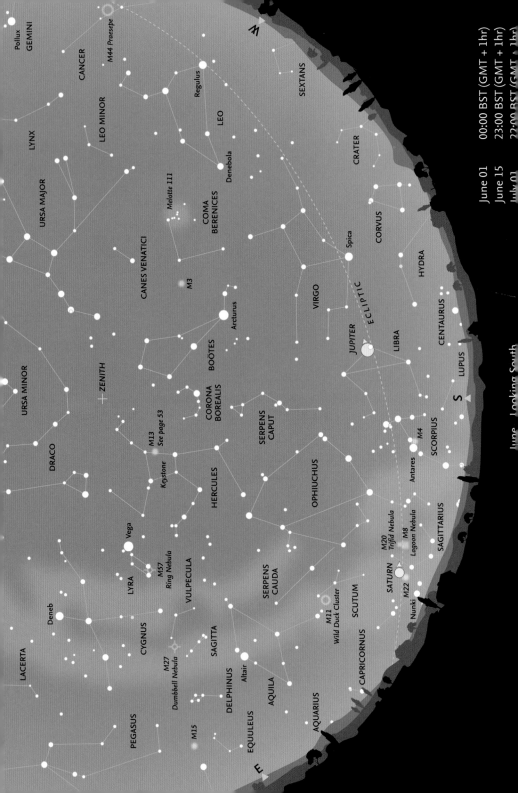

June – Looking South

June 01   00:00 BST (GMT + 1hr)
June 15   23:00 BST (GMT + 1hr)
July 01   22:00 BST (GMT + 1hr)

June – Looking South

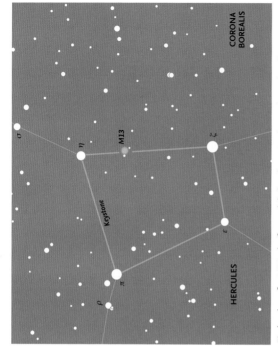

CORONA BOREALIS

HERCULES

Although the persistent twilight makes observing even the southern sky difficult, the rather undistinguished constellation of **Libra** lies almost due south. The red supergiant star **Antares** – the name means the 'Rival of Mars' – in **Scorpius** is visible slightly to the east of the meridian, but the 'tail' or 'sting' remains below the horizon. Higher in the sky is the large constellation of **Ophiuchus** (the 'Serpent Bearer'), lying between the two halves of the constellation of **Serpens: Serpens Caput** ('Head of the Serpent') to the west and **Serpens Cauda** ('Tail of the Serpent') to the east. (Serpens is the only constellation to be divided into two distinct parts.) The ecliptic runs across Ophiuchus, and the Sun actually spends far more time in the constellation than it does in the 'classical' zodiacal constellation of Scorpius, a small area of which lies between Libra and Ophiuchus.

Higher in the southern sky, the three constellations of **Boötes**, **Corona Borealis** and **Hercules** are now better placed for observation than at any other time of the year. This is an ideal time to observe the fine globular cluster of M13 in Hercules.

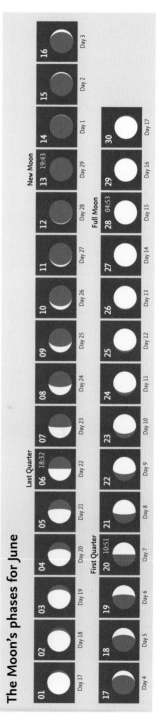

Finder chart for M13, the finest globular cluster in the northern sky. All stars down to magnitude 7.5 are shown.

The Moon's phases for June

Last Quarter
First Quarter
New Moon 13 19:43
Full Moon 28 04:53

01 Day 17	02 Day 18	03 Day 19	04 Day 20	05 Day 21	06 18:32 Day 22	07 Day 23
08 Day 24	09 Day 25	10 Day 26	11 Day 27	12 Day 28	13 19:43 Day 29	14 Day 1
15 Day 2	16 Day 3					
17 Day 4	18 Day 5	19 Day 6	20 10:51 Day 7	21 Day 8	22 Day 9	23 Day 10
24 Day 11	25 Day 12	26 Day 13	27 Day 14	28 04:53 Day 15	29 Day 16	30 Day 17

June – Moon and Planets

The Moon

On June 12, one day before New Moon, the waning crescent passes 1.2° north of **Aldebaran**, but there is no visible occultation. On June 18, as a waxing crescent, it passes **Regulus** in **Leo**, low in the western sky. On June 22, as waxing gibbous (two days after First Quarter) it is in **Virgo**, nor far from **Spica**. On June 25 (nearly Full) it passes close to **Antares** in **Scorpius**.

The planets

Mercury is behind the Sun at superior conjunction on June 6, but rapidly moves east of the Sun. **Venus** (mag. -4.0 to -4.1) is a prominent object in the evening sky, rapidly moving from **Gemini**, through **Cancer**

and just into **Leo** by the end of the month. **Mars** (mag. -1.2 to -2.1) is in **Capricornus**, rising about midnight early in the month. It slows its eastward motion and begins to retrograde right at the end of June. **Jupiter**, still in **Libra** (at mag. -2.5 to -2.3), continues slow retrograde motion. **Saturn** is retrograding slowly inside **Sagittarius**, and reaches opposition on June 27 at magnitude 0.0, unfortunately just before Full Moon. **Uranus** is still moving east in **Aries** at magnitude 5.9, hidden in the dawn sky. **Neptune**, in **Aquarius**, may be seen in the pre-dawn sky, and is magnitude 7.9 throughout the month. The minor planet **(4)** **Vesta** comes to opposition in **Sagittarius** (at mag. 5.3) on June 19. A chart showing both Saturn and Vesta is given on page 51.

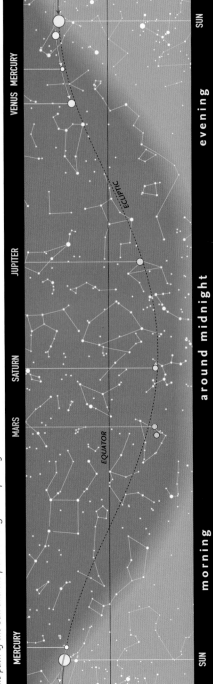

The path of the Sun and the planets along the ecliptic in June.

Calendar for June

Evening 22:00 (BST)

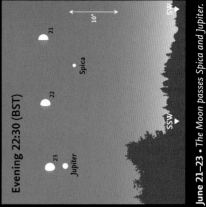

After midnight 1:00 (BST)

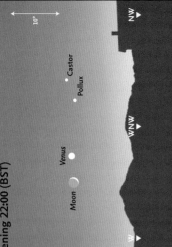

June 1 • *The Moon is close to Saturn in the south, shortly after midnight.*

June 16 • *The Moon with Venus and, a little more to the northwest, Castor and Pollux.*

Evening 22:30 (BST)

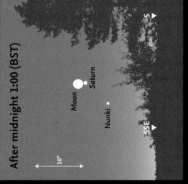

Evening 22:30 (BST)

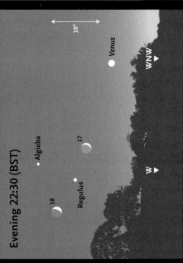

June 17–18 • *The Moon passes Regulus and Algieba. Venus is a little lower to the horizon.*

June 21–23 • *The Moon passes Spica and Jupiter.*

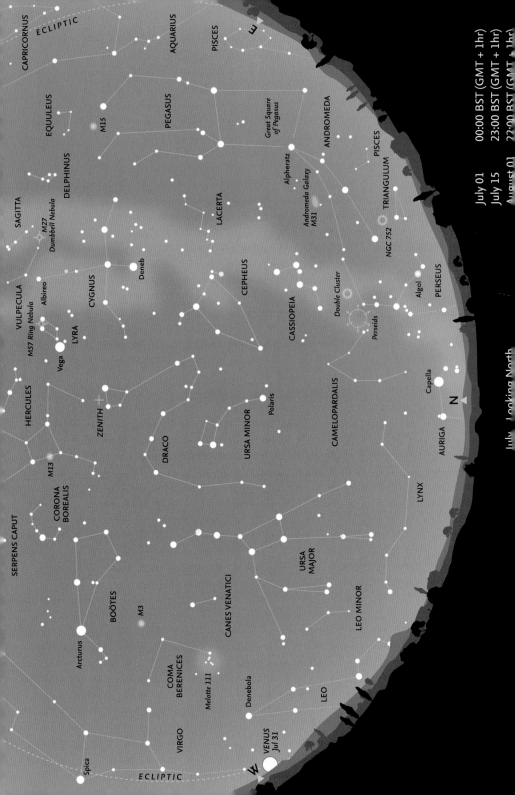

CAPRICORNUS
ECLIPTIC
AQUARIUS
PISCES
EQUULEUS
PEGASUS
M15
Great Square
of Pegasus
ANDROMEDA
PISCES
DELPHINUS
Alpheratz
Andromeda Galaxy
M31
TRIANGULUM
SAGITTA
M27
Dumbbell Nebula
LACERTA
NGC 752
Deneb
CEPHEUS
Double Cluster
PERSEUS
VULPECULA
Albireo
CYGNUS
CASSIOPEIA
Perseids
Algol
M57 Ring Nebula
LYRA
Capella
Vega
HERCULES
CAMELOPARDALIS
ZENITH
Polaris
AURIGA
N
CORONA
BOREALIS
DRACO
URSA MINOR
M13
LYNX
SERPENS CAPUT
BOÖTES
M3
CANES VENATICI
URSA
MAJOR
Arcturus
LEO MINOR
COMA
BERENICES
Melotte 111
Denebola
LEO
VIRGO
Spica
VENUS
Jul 31
ECLIPTIC
W
E
N

July 01   00:00 BST (GMT + 1hr)
July 15   23:00 BST (GMT + 1hr)
August 01   22:00 BST (GMT + 1hr)

July, Looking North

July – Looking North

As in June, light nights and the chance of observing noctilucent clouds persist throughout July, but later in the month (and particularly after midnight) some of the major constellations begin to be more easily seen. **Capella**, the brightest star in **Auriga** (most of which is too low to be visible), is skimming the northern horizon. **Cassiopeia** is clearly visible in the northeast and **Perseus**, to its south, is beginning to climb clear of the horizon. The band of the Milky Way, from Perseus through Cassiopeia towards **Cygnus**, stretches up into the northeastern sky. If the sky is dark and clear, you may be able to make out the small, faint constellation of **Lacerta**, lying across the Milky Way between Cassiopeia and Cygnus. In the east, the stars of **Pegasus** are now well clear of the horizon, with the main line of stars forming **Andromeda** roughly parallel to the horizon in the northeast. **Alpheratz** (α Andromedae) is actually the star at the northeastern corner of the **Great Square of Pegasus**. **Cepheus** and **Ursa Minor**, in the east and west, respectively. The head of **Draco** is very close to the zenith so the whole of this winding constellation is readily seen.

Meteors

July brings increasing meteor activity, mainly because there are several minor radiants active in the constellations of **Capricornus** and **Aquarius**. Because of their location, however, observing conditions are not particularly favourable for northern-hemisphere observers, although the first shower, the **Alpha Capricornids**, active from July 11 to August 10 (peaking July 27–28), does often produce very bright fireballs. The maximum rate, however, is only about 5 per hour. The parent body is Comet 169P/NEAT. The most prominent shower is probably that of the **Delta Aquariids**, which are active from around July 21 to August 23, with a peak on July 28–29, although even then the hourly rate is unlikely to reach 20 meteors per hour. In this case, the parent body is possibly

Comet 96P/Machholz. This year both shower maxima occur at, or just after Full Moon, so observing conditions are very unfavourable. The **Perseids**, begin on July 13 and peak on August 11–12, when there is a waxing gibbous Moon. A chart showing the **Delta Aquariid** radiant is shown on page 16.

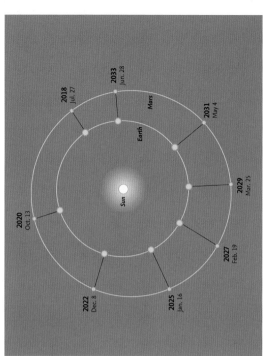

The opposition of Mars on July 27 is particularly favourable, because Mars is both close to the Earth and near its perihelion (closest to the Sun). Although opposition is at Full Moon, the planet will be readily visible some days before and after that date. A chart showing the location of Mars around opposition appears on page 59.

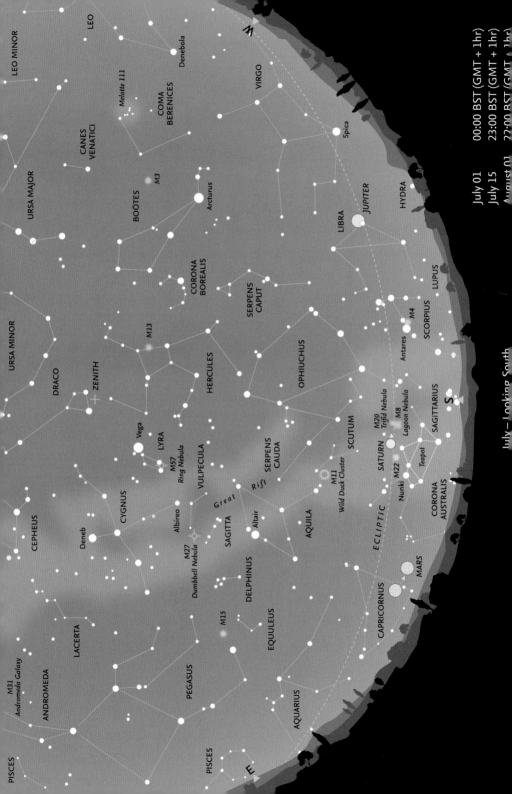

July – Looking South

July – Looking South

Although part of the constellation remains hidden, this is perhaps the best time of year to see **Scorpius**, with deep red **Antares** (α Scorpii), glowing just above the southern horizon. At around midnight (UT) 01:00 BST, part of **Sagittarius**, with the distinctive asterism of the 'Teapot', and the dense star clouds of the centre of the Milky Way, are just visible in the south. With clear skies, the Great Rift – actually dust clouds that hide the more distant stars – runs down the Milky Way from Cygnus towards Sagittarius. The sprawling constellation of **Ophiuchus** lies close to the meridian for a large part of the month, separating the two halves of the constellation of **Serpens**. The western half is called **Serpens Caput** (Head of the Serpent) and the eastern part **Serpens Cauda** (Tail of the Serpent). In the east, the bright **Summer Triangle**, consisting of **Vega** in **Lyra**, **Deneb** in **Cygnus** and **Altair** in **Aquila**, begins to dominate the southern sky, as it will throughout August and into September. The small constellation of Lyra, with Vega and a distinctive quadrilateral of stars to its east and south, lies not far south of the zenith.

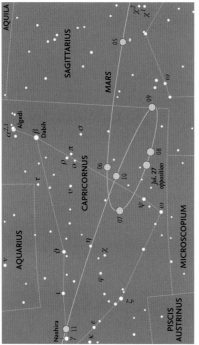

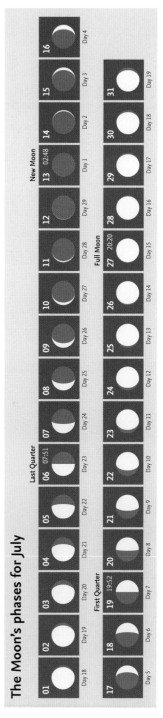

Finder chart for Mars around the time of opposition on July 27. Background stars are shown down to magnitude 6.5.

The Moon's phases for July

01 — Day 18	02 — Day 19	03 — Day 20
04 — First Quarter — Day 21	05 — Day 22	06 — Last Quarter 07:51 — Day 23
07 — Day 24	08 — Day 25	09 — Day 26
10 — Day 27	11 — Day 28	12 — Day 29
13 — New Moon 02:48 — Day 1	14 — Day 2	15 — Day 3
16 — Day 4	17 — Day 5	18 — Day 6
19 — First Quarter 19:52 — Day 7	20 — Day 8	21 — Day 9
22 — Day 10	23 — Day 11	24 — Day 12
25 — Day 13	26 — Day 14	27 — Full Moon 20:20 — Day 15
28 — Day 16	29 — Day 17	30 — Day 18
31 — Day 19		

July – Moon and Planets

The Moon

On July 13 at New Moon, there is a partial solar eclipse, visible from the Southern Ocean only, south of Australia. On July 10 the Moon passes *Aldebaran*, on July 13, *Pollux*, and on July 15, *Regulus*, all in daylight. On July 23 it is north of *Antares* in *Scorpius*. Full Moon is on July 27. On that date there is a total lunar eclipse, visible from India, Africa and Europe.

The Earth

The Earth passes aphelion (the farthest point from the Sun in its orbit) on July 6 at 16:47 UT, when its distance is 1.01670 AU (152,095,557 km).

The planets

Mercury reaches greatest eastern elongation on July 12 (26.4°E, mag. 0.4), but it is far too low on the western horizon to be visible. *Venus* is very bright (mag. -4.1 to -4.3) and may just be visible in the evening sky just before it sets. *Mars* is retrograding in *Capricornus*, and comes to perihelic opposition on July 27, unfortunately close to the Full Moon. *Jupiter* is moving very slowly in *Libra*, initially at magnitude -2.3, reaching a stationary point on July 11 and then resumes direct motion. It fades slightly to mag. -2.1 at the end of the month. *Saturn* is retrograding in *Sagittarius* and fades slightly from magnitude 0.0 to 0.2. *Uranus* (magnitude 5.8), is moving very slowly east just inside *Aries*. *Neptune* is still in *Aquarius*, retrograding slowly and at magnitude 7.9–7.8.

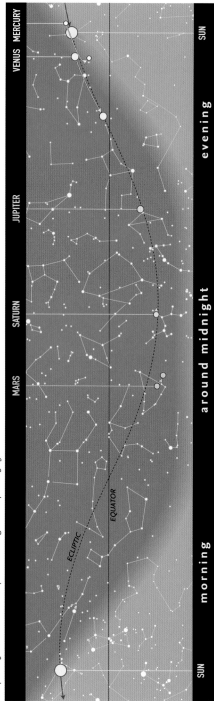

The path of the Sun and the planets along the ecliptic in July.

1	01:43	Mars 4.8°S of Moon
6	07:51	Last Quarter
6	16:47	Earth at aphelion
		(152,095,557 km = 1.016696 AU)
9	20:00*	Regulus 1.1°S of Venus
0	09:55	Aldebaran 1.1°S of Moon
Jul.11–Aug.10		Alpha Capricornid meteor shower
2	05:29	Mercury greatest elongation
		(26.4°E, mag. 0.4)
3	02:48	New Moon
3	03:01	Partial solar eclipse
		(Southern Ocean only)
3	08:25	Moon at perigee (357,400 km)
3	09:23	Pollux 7.8°N of Moon
Jul.13–Aug.26		Perseid meteor shower
5	16:38	Regulus 1.8°S of Moon
6	03:31	Venus 1.6°S of Moon
9	09:12	Spica 7.7°S of Moon
9	19:52	First Quarter
0	23:57	Jupiter 4.6°S of Moon
Jul.21–Aug.23		Delta Aquariid meteor shower
3	03:09	Antares 9.1°S of Moon
5	05:43	Saturn 2.0°S of Moon
7	05:13	Mars at opposition (mag. -2.8)
7	05:44	Moon at apogee (406,200 km)
7	20:20	Full Moon
7	20:22	Total lunar eclipse
		(India, Africa, Europe)
7	22:05	Mars 6.7°S of Moon
7–28		Alpha Capricornid shower maximum
8–29		Delta Aquariid shower maximum

These objects are close together for an extended

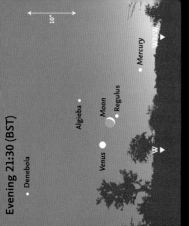

Evening 21:45 (BST)

Algieba ·
Regulus · Venus
· Mercury
W ▶
WNW ▶
10°

July 9 • Venus is close to Regulus. Mercury is closer to the horizon but may not be easy to detect.

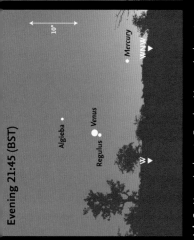

Evening 23:00 (BST)

Moon
Jupiter ·
Spica ·
SW ▶
WSW ▶
10°

July 20 • The Moon with Jupiter and, close to the horizon, Spica.

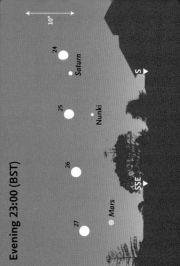

Evening 21:30 (BST)

· Denebola
Algieba ·
Venus
Moon
Regulus
· Mercury
W ▶
WNW ▶
10°

July 15 • The Moon joins Regulus and Venus. Mercury remains close to the horizon.

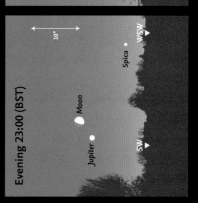

Evening 23:00 (BST)

24
25 Saturn
26 Nunki
27
Mars
S ▶
SSE ▶
10°

July 24–27 • The Moon passes Saturn and Mars on the day of its opposition. Mars is now magnitude -2.8.

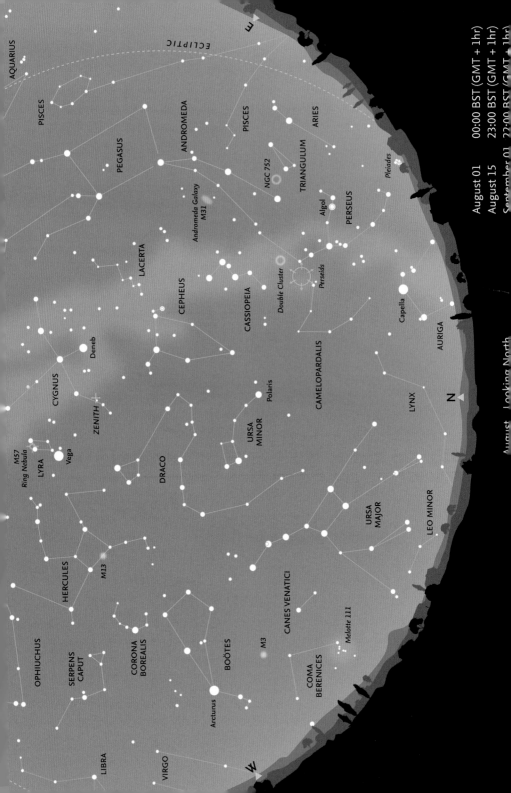

AQUARIUS

PISCES

ECLIPTIC

PEGASUS

ANDROMEDA

PISCES

ARIES

NGC 752

Andromeda Galaxy
M31

TRIANGULUM

LACERTA

CEPHEUS

Double Cluster

Perseids

PERSEUS

Algol

CASSIOPEIA

Pleiades

Deneb

Capella

CYGNUS

CAMELOPARDALIS

AURIGA

ZENITH

Polaris

N

DRACO

URSA
MINOR

LYNX

Vega

LYRA

Ring Nebula
M57

URSA
MAJOR

LEO MINOR

HERCULES

M13

OPHIUCHUS

CANES VENATICI

SERPENS
CAPUT

Melotte 111

CORONA
BOREALIS

COMA
BERENICES

M3

BOÖTES

Arcturus

LIBRA

VIRGO

W

August 01   00:00 BST (GMT + 1hr)
August 15   23:00 BST (GMT + 1hr)
September 01 22:00 BST (GMT + 1hr)

August – Looking North

August – Looking North

Ursa Major is now the 'right way up' in the northwest, although some of the fainter stars in the south of the constellation are difficult to see. Beyond it, *Boötes* stands almost vertically in the west, but pale orange *Arcturus* is sinking towards the horizon. Higher in the sky, both *Corona Borealis* and *Hercules* are clearly visible.

In the northeast, *Capella* is clearly seen, but most of *Auriga* still remains below the horizon. Higher in the sky, *Perseus* is gradually coming into full view and, later in the night and later in the month, the beautiful *Pleiades* cluster rises above the northeastern horizon. Between Perseus and *Polaris* lies the faint and unremarkable constellation of *Camelopardalis*.

Higher still, both *Cassiopeia* and *Cepheus* are well placed for observation, despite the fact that Cassiopeia is completely immersed in the band of the Milky Way, as is the 'base' of Cepheus. *Pegasus* and *Andromeda* are now well above the eastern horizon and, below them, the constellation of *Pisces* is climbing into view. Two of the stars in the *Summer Triangle*, *Deneb* and *Vega*, are close to the zenith high overhead.

Meteors

August is the month when one of the best meteor showers of the year occurs: the *Perseids*. This is a long shower, generally beginning about July 13 and continuing until around August 26, with a maximum in 2018 on August 11–12, when the rate may reach as high as 100 meteors per hour (and on rare occasions, even higher). In 2018, maximum is on the night of New Moon, so conditions are especially favourable. The Perseids are debris from Comet 109P/Swift-Tuttle (the Great Comet of 1862). Perseid meteors are fast and many of the brighter ones leave persistent trains. Some bright fireballs also occur during the shower.

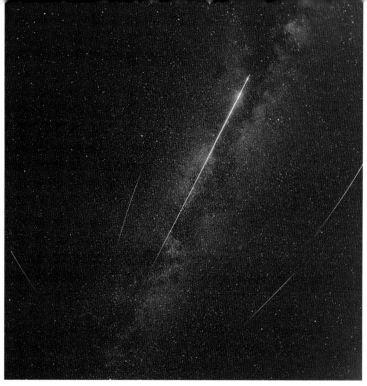

A brilliant Perseid fireball, streaking alongside the Great Rift in the Milky Way, photographed in 2012 by Jens Hackmann from near Weikersheim in Germany. Four additional, fainter Perseids are also visible in the image.

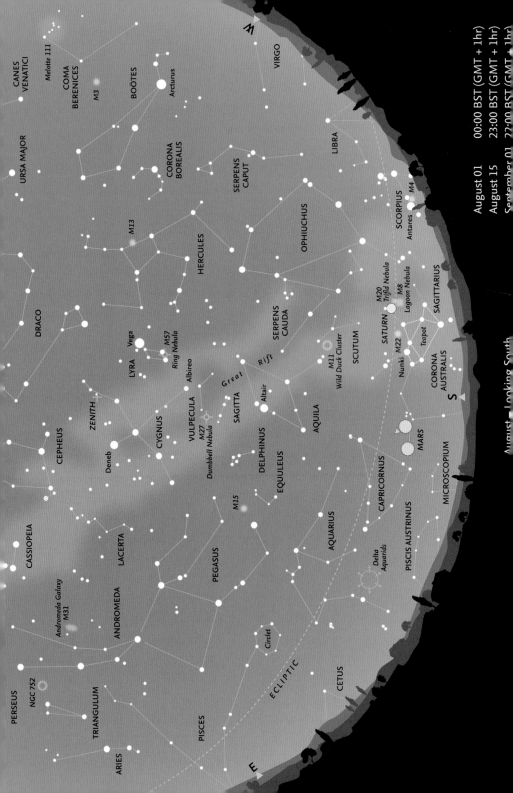

August – Looking South

The whole stretch of the summer Milky Way stretches across the sky in the south, from **Cygnus**, high in the sky near the zenith, past **Aquila**, with bright **Altair** (α Aquilae), to part of the constellation of **Sagittarius** close to the horizon, where the pattern of stars known as the 'Teapot' may be just visible. This area contains many nebulae and both open and globular clusters. Between **Albireo** (β Cygni) and Altair lie the two small constellations of **Vulpecula** and **Sagitta**, with the latter easier to distinguish (because of its shape) from the clouds of the Milky Way. Between Sagitta and **Pegasus** to the east lie the highly distinctive five stars that form the tiny constellation of **Delphinus** (again, one of the few constellations that actually bear some resemblance to the creatures after which they are named). Below Aquila, mainly in the star clouds of the Milky Way, lies **Scutum**, most famous for the bright open cluster, M11 or the 'Wild Duck Cluster', readily visible in binoculars. To the southeast of Aquila lie the two zodiacal constellations of **Capricornus** and **Aquarius**.

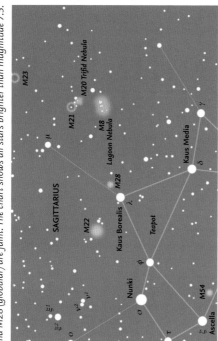

A finder chart for the gaseous nebulae M8 (the Lagoon Nebula) and M20 (the Trifid Nebula) and the globular cluster M22, all in Sagittarius. Clusters M21 & M23 (open) and M28 (globular) are faint. The chart shows all stars brighter than magnitude 7.5.

The Moon's phases for August

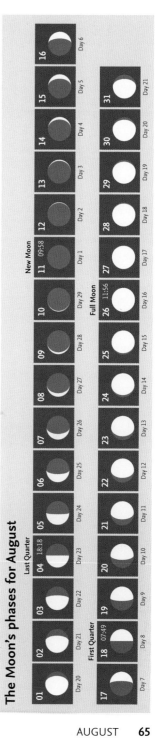

August – Moon and Planets

The Moon

The Moon passes close to **Aldebaran** in **Taurus** on August 6. Part of an occultation is visible from eastern Asia. On August 11, at New Moon, there is a partial solar eclipse, visible from northern Asia and Europe (including Scandinavia). On August 12, the Moon passes north of **Regulus** in **Leo**, but there is no occultation and the event is too close to the Sun to be visible. On August 18, it may be possible to catch a glimpse of the waxing gibbous Moon not far from **Antares** in **Scorpius** in the evening sky, just before the objects set in the west.

The planets

Mercury passes inferior conjunction (between Earth and the Sun), and is thus invisible, on August 9. It reaches greatest western elongation (18.3°W, mag. -0.2) on August 26, and might be glimpsed with difficulty in the east just before sunrise. On August 17, **Venus** (at magnitude -4.5) reaches greatest eastern elongation in **Virgo**, but is very low in the west just before it sets. Following opposition on July 27, **Mars** continues to retrograde in **Capricornus**, reaching a stationary point on August 28, just crossing into **Sagittarius**, but then resumes direct motion. It fades from mag. -2.8 to -2.1 over the month. **Jupiter** (mag. -2.1 to -1.9) is moving slowly eastwards in **Libra** and may be glimpsed, low in the west, early in the month, shortly after sunset. **Saturn** (magnitude 0.2–0.3) is in **Sagittarius** throughout August, moving extremely slowly towards the west. **Uranus** (magnitude 5.8–5.7), reaches a stationary point in **Aries** on August 8 and begins retrograde motion. **Neptune** (magnitude 7.8) is farther west, in **Aquarius.**

The path of the Sun and the planets along the ecliptic in August.

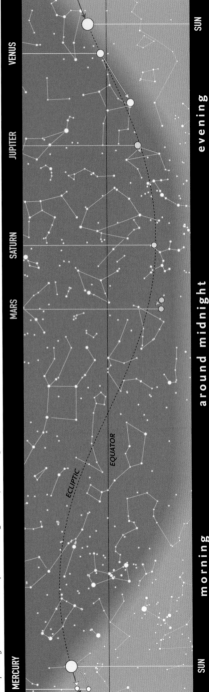

Calendar for August

04	18:18	Last Quarter
06	19:01	Aldebaran 1.1°S of Moon
09	02:06	Mercury inferior conjunction
09	20:07	Pollux 7.8°N of Moon
10	18:07	Moon at perigee (358,100 km)
11	01:31	Mercury 5.5°S of Moon
11	09:46	Partial solar eclipse (Northern Asia & Europe)
11	09:58	New Moon
11–12		Perseid meteor shower maximum
12	03:07	Regulus 1.8°S of Moon
14	13:34	Venus 6.3°S of Moon
15	17:21	Spica 7.7°S of Moon
17	10:38	Jupiter 4.5°S of Moon
17	17:31	Venus greatest elongation (45.9°E, mag. −4.5)
18	07:49	First Quarter
19	09:31	Antares 9.1°S of Moon
21	09:38	Saturn 2.1°S of Moon
23	11:23	Moon at apogee (405,700 km)
23	17:14	Mars 6.8°S of Moon
26	11:56	Full Moon
26	20:34	Mercury greatest elongation (18.3°W, mag. −0.2)
28		Alpha Aurigid first maximum

Morning 5:00 (BST)

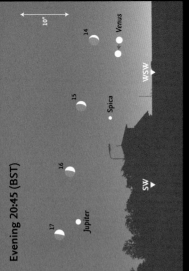

August 6–7 • The Moon with Aldebaran. Elnath (β Tau) and the Pleiades are nearby.

Evening 20:45 (BST)

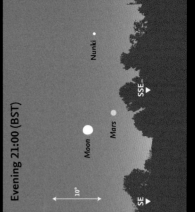

August 14 –17 • The Moon passes Venus, Spica and Jupiter, low in the southwestern sky.

Evening 21:00 (BST)

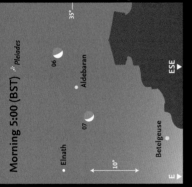

August 20–21 • The Moon passes Saturn. Antares, Sabik (η Oph) and Nunki (σ Sgr) are nearby.

Evening 21:00 (BST)

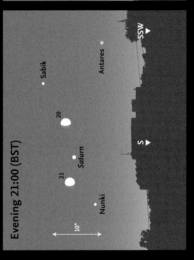

August 23 • The Moon with Mars, low in the southeast.

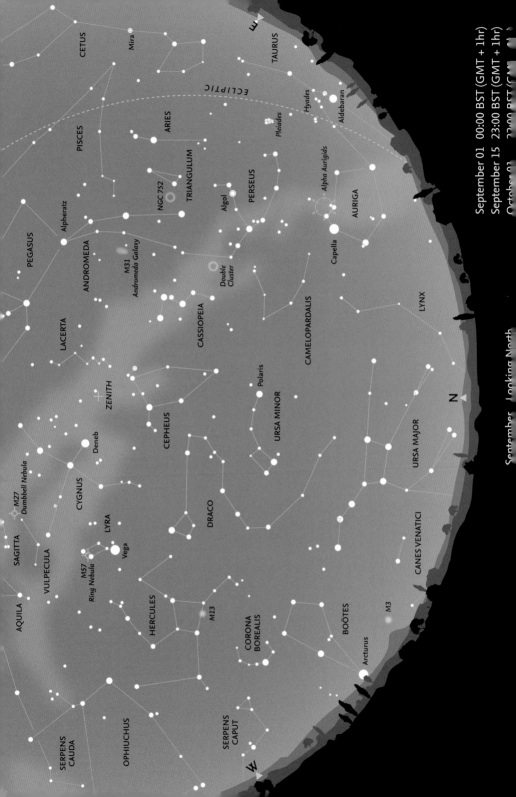

September 01 00:00 BST (GMT + 1hr)
September 15 23:00 BST (GMT + 1hr)
October 01 22:00 BST (GMT

September Looking North

September – Looking North

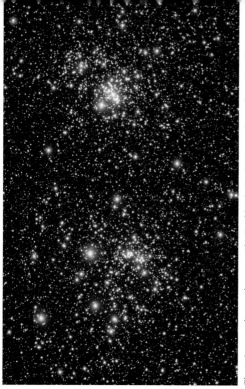

The twin open clusters, known as the Double Cluster, in Perseus (more formally called h and χ Persei) are close to the main portion of the Milky Way.

Ursa Major is now low in the north and to the northwest **Arcturus** and much of **Boötes** sink below the horizon later in the night and later in the month. In the northeast **Auriga** is beginning to climb higher in the sky. Later in the month, **Taurus**, with orange **Aldebaran** (α Tauri), and even **Gemini**, with **Castor** and **Pollux**, become visible in the east and northeast. Due east, **Andromeda** is now clearly visible, with the small constellations of **Triangulum** and **Aries** (the latter a zodiacal constellation) directly below it. Practically the whole of the Milky Way is visible, arching across the sky, both in the north and in the south. It is not particularly clear in Auriga, or even **Perseus**, but in **Cassiopeia** and on towards **Cygnus** the clouds of stars become easier to see. **Cepheus** is 'upside-down' near the zenith, and the head of **Draco** and **Hercules** beyond it are well placed for observation.

Meteors

After the major Perseid shower in August, there is very little shower activity in September. One minor, but very extended, shower, known as the **Alpha Aurigids**, actually has two peaks of activity. The first was on August 28, but the primary peak occurs on September 15. At either of the maxima, however, the hourly rate hardly reaches 10 meteors per hour, although the meteors are bright and relatively easy to photograph. Activity from this shower also extends into October. The **Southern Taurid** shower begins this month and, although rates are low, often produces very bright fireballs. As a slight compensation for the lack of activity, however, in September the number of sporadic meteors reaches its highest rate than at any other time during the year.

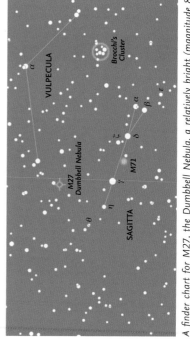

A finder chart for M27, the Dumbbell Nebula, a relatively bright (magnitude 8) planetary nebula – a shell of material ejected in the late stages of a star's lifetime – in the constellation of Vulpecula. All stars brighter than magnitude 7.5 are shown.

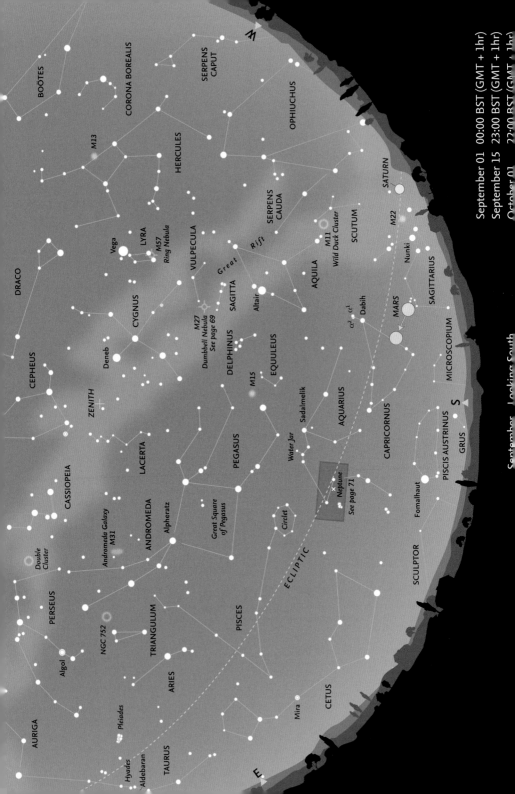

E
S
N

BOÖTES
CORONA BOREALIS
SERPENS CAPUT
OPHIUCHUS
DRACO
M13
HERCULES
SATURN
SERPENS CAUDA
SCUTUM
M11
Wild Duck Cluster
M22
Nunki
AQUILA
SAGITTARIUS
Vega
LYRA
M57
Ring Nebula
VULPECULA
Great Rift
CYGNUS
Deneb
M27
Dumbbell Nebula
See page 69
SAGITTA
Altair
Dabih
α² α¹
MARS
CEPHEUS
ZENITH
LACERTA
DELPHINUS
EQUULEUS
M15
MICROSCOPIUM
CASSIOPEIA
Andromeda Galaxy
M31
ANDROMEDA
Alpheratz
PEGASUS
Great Square
of Pegasus
Sadalmelik
Water Jar
AQUARIUS
CAPRICORNUS
PISCIS AUSTRINUS
GRUS
Double
Cluster
NGC 752
Circlet
Neptune
See page 71
SCULPTOR
Fomalhaut
PERSEUS
Algol
TRIANGULUM
PISCES
ECLIPTIC
CETUS
Mira
ARIES
AURIGA
Pleiades
Hyades
Aldebaran
TAURUS

September – Looking South

The **Summer Triangle** is now high in the southwest, with the Great Square of **Pegasus** high in the southeast. Below Pegasus are the two zodiacal constellations of **Capricornus** and **Aquarius**. In what is otherwise an unremarkable constellation, α Capricorni (**Algedi**) is actually a visual binary, with the two stars (α¹ Cap and α² Cap) readily seen with the naked eye. **Dabih** (β Capricorni), just to the south, is also a double star, and the components are relatively easy to separate with binoculars. In Aquarius, just to the east of **Sadalmelik** (α Aquarii) there is a small asterism consisting of four stars, resembling a tiny letter 'Y', known as the 'Water Jar'. Below Aquarius is a sparsely populated area of the sky with just one bright star in the constellation of **Piscis Austrinus**. In classical illustrations, water is shown flowing from the 'Water Jar' towards bright **Fomalhaut** (α Piscis Austrini).

Another zodiacal constellation, **Pisces**, is now clearly visible to the east of Aquarius. Although faint, there is a distinctive asterism of stars, known as the 'Circlet', south of the Great Square and another l ne of faint stars to the east of Pegasus. Still farther down towards

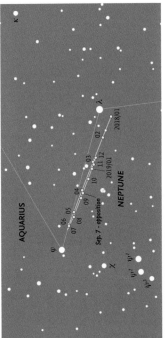

The path of Neptune in 2018. Neptune comes to opposition on September 7. All stars brighter than magnitude 8.5 are shown.

the horizon is the constellation of **Cetus**, with the famous variable star **Mira** (o Ceti) at its centre. When Mira is at maximum brightness (around mag. 3.5) it is clearly visible to the naked eye, but it disappears as it fades towards minimum (about mag. 9.5 or less). There is a finder chart for Mira on page 83.

The Moon's phases for September

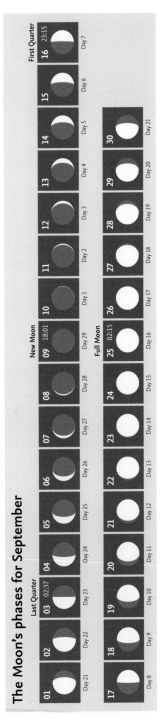

September – Moon and Planets

The Moon

On September 3, the Last Quarter Moon passes very close to *Aldebaran* in *Taurus*, but no occultation is visible. On September 6 the waning crescent is in *Gemini*, below *Castor* and *Pollux*. On September 8, one day before New Moon, it passes close to *Regulus* and *Mercury* in daylight. On September 12 it is north of *Spica* in *Virgo*. On September 15, just before First Quarter, it passes north of *Antares*.

The planets

Mercury passes superior conjunction on the far side of the Sun on September 21. Early in the month *Venus* (brilliant at magnitude -4.6) is in *Virgo*, close to *Spica* in the evening sky, but moves to the border of *Libra*, close to the boundary of *Hydra* by the end of the month.

Mars fades from mag. -2.1 to -1.3 over the course of the month as it continues direct motion in *Capricornus*. *Jupiter* (mag. -1.9 to -1.8) moves eastwards in *Libra* in the evening sky. *Saturn* remains in *Sagittarius* at magnitude 0.4 to 0.5, but the constellation is low, close to the horizon. *Uranus* (magnitude 5.7) is retrograding very slowly in *Aries*, as is *Neptune* in *Aquarius*. Neptune comes to opposition on September 7 at magnitude 7.8, not far from φ Aquarii. This is two days before New Moon (on September 9) so the waning crescent (in *Cancer*) should not cause interference. The planet should be visible around a week before and after opposition.

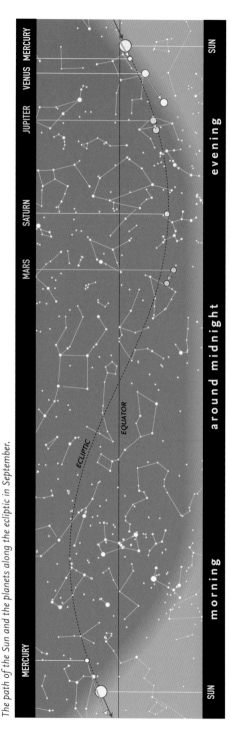

The path of the Sun and the planets along the ecliptic in September.

Calendar for September

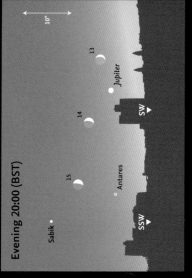

After midnight 2:00 (BST)

September 3 • The Moon with Aldebaran. Elnath (β Tau) is to the left.

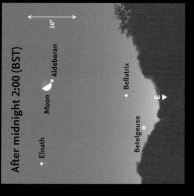

Evening 17:30 (BST)

September 17–19 • The Moon passes Saturn, Nunki (σ Sgr) and Mars.

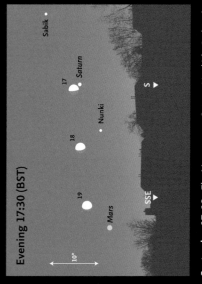

Evening 20:00 (BST)

September 13–15 • The Moon passes Jupiter between September 13 and 14. One day later it is close to Antares and Sabik (η Oph).

Morning 5:30 (BST)

September 30 • The Moon is again close to Aldebaran. Betelgeuse is near.

02	08:00*	Spica 1.4°N of Venus
03	02:01	Aldebaran 1.2°S of Moon
03	02:37	Last Quarter
06	05:29	Pollux 7.8°N of Moon
07	18:27	Neptune at opposition (mag. 7.8)
Sep.07–Nov.19		Southern Taurid meteor shower
08	01:20	Moon at perigee (361,400 km)
08	13:37	Regulus 1.8°S of Moon
08	22:16	Mercury 0.9°S of Moon
09	18:01	New Moon
12	03:08	Spica 7.6°S of Moon
12	15:46	Venus 10.7°S of Moon
14	02:20	Jupiter 4.4°S of Moon
15	17:14	Antares 9.0°S of Moon
15		Alpha Aurigid second maximum
16	23:15	First Quarter
17	16:30	Saturn 2.1°S of Moon
20	00:54	Moon at apogee (404,900 km)
20	06:42	Mars 4.8°S of Moon
21	01:52	Mercury superior conjunction
23	01:54	Autumnal equinox
25	02:52	Full Moon
27	03:00*	Vesta 2.8°S of Saturn
30	07:33	Aldebaran 1.4°S of Moon

* These objects are close together for an extended period around this time.

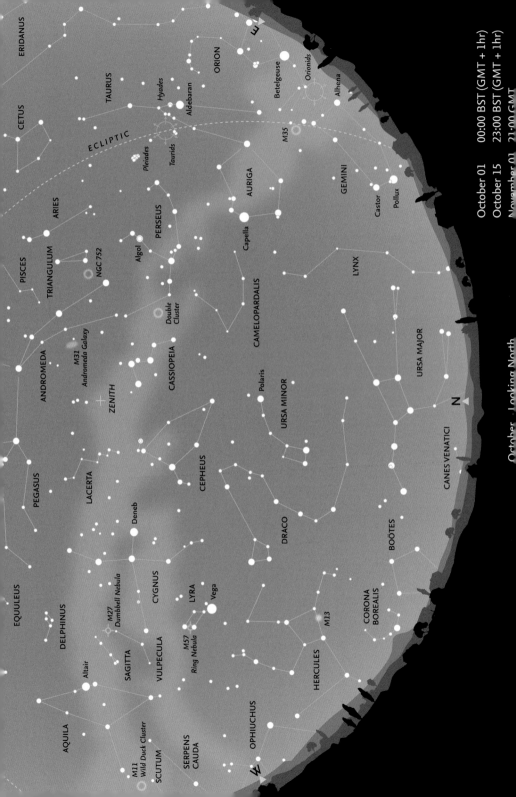

October – Looking North

October 01 — 00:00 BST (GMT + 1hr)
October 15 — 23:00 BST (GMT + 1hr)
November 01 — 21:00 GMT

October – Looking North

Ursa Major is grazing the horizon in the north, while high overhead are the constellations of **Cepheus**, **Cassiopeia** and **Perseus**, with the Milky Way between Cepheus and Cassiopeia at the zenith. **Auriga** is now clearly visible in the east, as is **Taurus** with the **Pleiades**, **Hyades** and orange **Aldebaran**. Also in the east, **Orion** and **Gemini** are starting to rise clear of the horizon.

The constellations of **Boötes** and **Corona Borealis** are now essentially lost to view in the northwest, and **Hercules** is also descending towards the western horizon. The three stars of the Summer Triangle are still clearly visible, although **Aquila** and **Altair** are beginning to approach the horizon in the west. Towards the end of the month (October 28) Summer Time ends in Europe with Britain reverting to Greenwich Mean Time and Europe to Central European Time.

Meteors

The **Orionids** are the major, fairly reliable meteor shower active in October. Like the May **Eta Aquariid** shower, the Orionids are associated with Comet 1P/Halley. During this second pass through the stream of particles from the comet, slightly fewer meteors are seen than in May, but conditions are more favourable for northern observers. In both showers the meteors are very fast, and many leave persistent trains. Although the Orionid maximum is quoted as October 21–22, in fact there is a very broad maximum, lasting about a week from October 20 to 27, with hourly rates around 25. Occasionally rates are higher (50–70 per hour). Unfortunately, in 2018, even the broad maximum extends from just before to just after Full Moon (October 24), so moonlight will cause considerable interference.

The faint shower of the **Southern Taurids** (often with bright fireballs) peaks on November 5–6. The Southern Taurid maximum occurs around the time of Full Moon, so conditions are unfavourable. Towards the end of the month, another shower (the **Northern Taurids**) begins to show activity, which peaks in mid-November. The parent comet for both Taurid showers is Comet 2P/Encke.

The path of Uranus in 2018. Uranus comes to opposition on October 24. All stars brighter than magnitude 7.5 are shown.

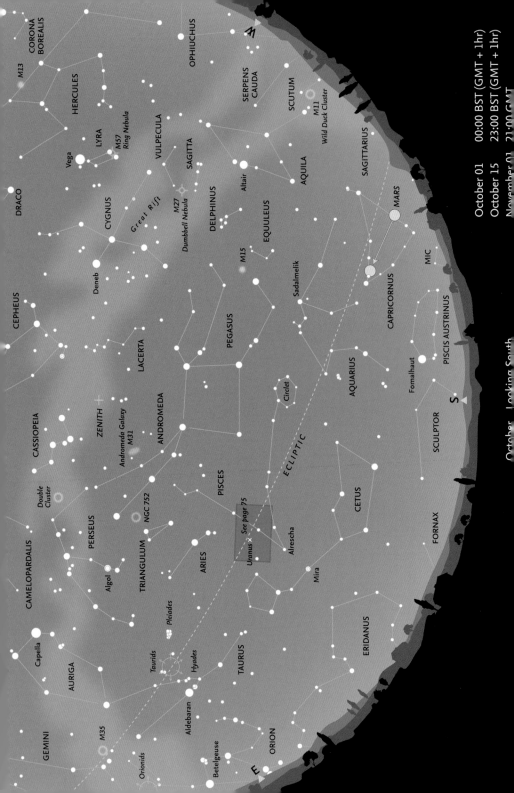

CORONA
BOREALIS

M13

HERCULES

DRACO

CEPHEUS

OPHIUCHUS

SERPENS
CAUDA

SCUTUM

M11
Wild Duck Cluster

SAGITTARIUS

LYRA

M57
Ring Nebula

Vega

VULPECULA

SAGITTA

CYGNUS

Great Rift

M27
Dumbbell Nebula

Deneb

AQUILA

Altair

DELPHINUS

MARS

EQUULEUS

CAMELOPARDALIS

LACERTA

M15

Sadalmelik

MIC

CAPRICORNUS

ZENITH

Andromeda Galaxy
M31

ANDROMEDA

PEGASUS

CASSIOPEIA

Double
Cluster

PERSEUS

NGC 752

Circlet

AQUARIUS

PISCIS AUSTRINUS

Algol

TRIANGULUM

PISCES

ECLIPTIC

Fomalhaut

Capella

AURIGA

ARIES

See page 25

Uranus

Alrescha

CETUS

S

SCULPTOR

GEMINI

M35

Orionids

Pleiades

Taurids

Hyades

TAURUS

Mira

FORNAX

Aldebaran

ERIDANUS

Betelgeuse

ORION

E

October 01   00:00 BST (GMT + 1hr)
October 15   23:00 BST (GMT + 1hr)
November 01   21:00 GMT

October Looking South

October – Looking South

The Great Square of **Pegasus** dominates the southern sky, framed by the two chains of stars that form the constellation of **Pisces**, together with **Alrescha** (α Piscium) at the point where the two lines of stars join. Also clearly visible is the constellation of **Cetus**, below Pegasus and Pisces. Although **Capricornus** is beginning to disappear, **Aquarius** to its east is well placed in the south, with solitary **Fomalhaut** and the constellation of **Piscis Austrinus** beneath it, close to the horizon.

The main band of the Milky Way and the Great Rift runs down from **Cygnus**, through **Vulpecula**, **Sagitta** and **Aquila** towards the western horizon. **Delphinus** and the tiny, unremarkable constellation of **Equuleus** lie between the band of the Milky Way and Pegasus. **Andromeda** is clearly visible high in the sky to the southeast, with the small constellation of **Triangulum** and the zodiacal constellation of **Aries** below it. **Perseus** is high in the east, and by now the **Pleiades** and **Taurus** are well clear of the horizon. Later in the night, and later in the month, **Orion** rises in the east, a sign that the autumn season has arrived and of the steady approach of winter.

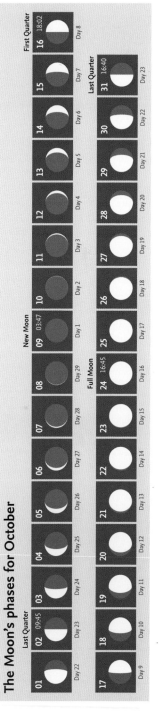

The constellation of Aquarius is one of the constellations that is visible in late summer and early autumn. The four stars forming the 'Y'-shape of the 'Water Jar' may be seen to the east of Sadalmelik (α Capricorni), the brightest star (top centre).

The Moon's phases for October

Last Quarter

01	02 09:45	03	04	05	06	07	08	09 03:47 New Moon	10	11	12	13	14	15	16 18:02 First Quarter
Day 9	Day 10	Day 11	Day 12	Day 13	Day 14	Day 15	Day 16	Day 1	Day 2	Day 3	Day 4	Day 5	Day 6	Day 7	Day 8

17	18	19	20	21	22	23	24 16:45 Full Moon	25	26	27	28	29	30	31 16:40 Last Quarter
Day 22	Day 23	Day 24	Day 25	Day 26	Day 27	Day 28	Day 29	Day 17	Day 18	Day 19	Day 20	Day 21	Day 22	Day 23

October – Moon and Planets

The Moon

On October 5, in the early morning, the Moon passes north of *Regulus* in *Leo*, before rising in the east. On October 9, at New Moon, it is close to *Spica* (and the Sun) in *Virgo*. On October 13, the waxing crescent is in *Ophiuchus*, north of *Antares* in *Scorpius*. On October 27, the waning gibbous Moon is in *Taurus*, passing south of Aldebaran in the daylight sky.

The planets

Mercury races east in October, but is far too low in daylight to be visible. *Venus*, moving westwards is lost in the daylight sky, south of the Sun. *Mars* (fading from mag. -1.3 to -0.6 over the month), is moving eastwards across *Capricornus*. *Jupiter* is moving eastwards in *Libra*, but is too close to the Sun to be visible. *Saturn* (at magnitude 0.5) is slowly moving eastward in *Sagittarius*. *Uranus* comes to opposition just within *Aries* at magnitude 5.7 on October 24 and continues retrograding as shown in the chart on page 75. *Neptune*, following opposition on September 7, continues retrograde motion in Aquarius at magnitude 7.8 throughout the month.

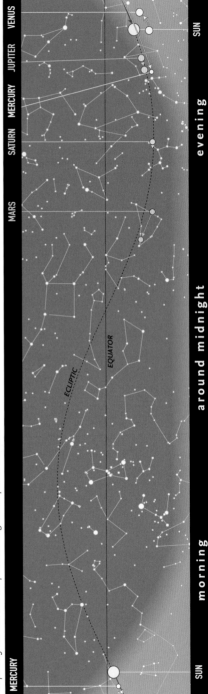

The path of the Sun and the planets along the ecliptic in October.

MERCURY   SUN   MARS   SATURN   MERCURY   JUPITER   VENUS   SUN

ECLIPTIC   EQUATOR

morning   around midnight   evening

Calendar for October

02	09:45	Last Quarter
03	12:28	Pollux 7.6°N of Moon
Oct.04–Nov.14		Orionid meteor shower
05	22:22	Regulus 1.9°S of Moon
05	22:27	Moon at perigee (366,400 km)
09–10		Southern Taurid maximum
09	03:47	New Moon
09	13:10	Spica 7.5°S of Moon
10	00:34	Mercury 5.9°S of Moon
10	14:49	Venus 13.2°S of Moon
11	21:20	Jupiter 4.1°S of Moon
13	02:08	Antares 8.7°S of Moon
14	15:00*	Mercury 6.8°N of Venus
15	02:46	Saturn 1.8°S of Moon
16	18:02	First Quarter
17	19:16	Moon at apogee (404,200 km)
18	13:02	Mars 2.0°S of Moon
Oct.19–Dec.10		Northern Taurid meteor shower
21–22		Orionid meteor shower maximum
24	00:47	Uranus at opposition (mag. 5.7)
24	16:45	Full Moon
26	14:16	Venus inferior conjunction
27	13:31	Aldebaran 1.6°S of Moon
28		Summer Time ends (Europe)
30	17:48	Pollux 7.3°N of Moon
31	16:40	Last Quarter
31	20:23	Moon at perigee (370,200 km)

These objects are close together for an extended period around this time.

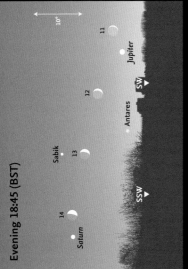

Morning 6:00 (BST)

October 5–6 • A narrow crescent Moon passes Regulus and Algieba, before sunrise.

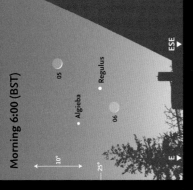

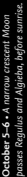

Evening 20:00 (BST)

October 17–18 • The Moon passes Mars. On October 17 Algedi and Dabih (α and β Cap resp.) are above the Moon.

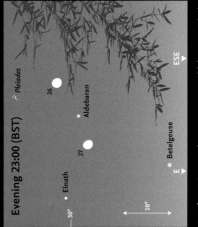

Evening 18:45 (BST)

October 11–14 • The Moon moves from Jupiter to Saturn and passes Antares, very close to the horizon, and Sabik.

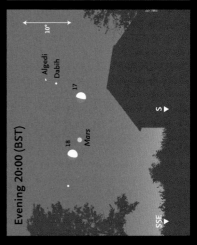

Evening 23:00 (BST)

October 26–27 • The Moon passes Aldebaran, with the Pleiades to the north.

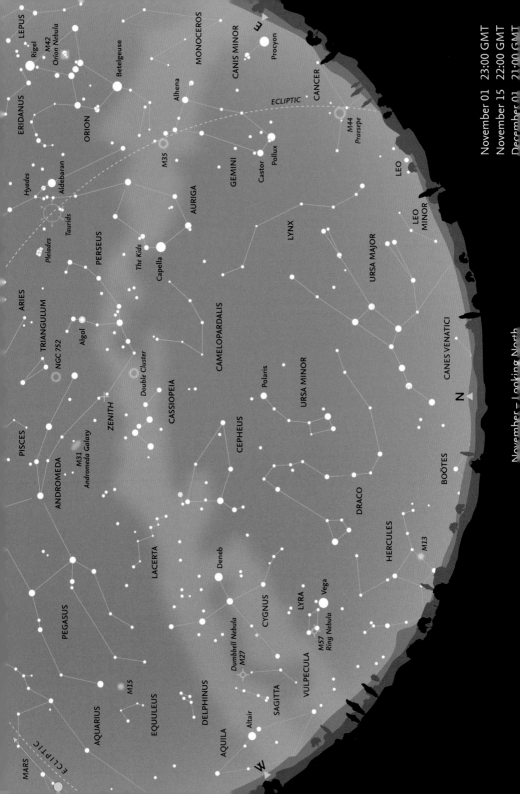

November 01 23:00 GMT
November 15 22:00 GMT
December 01 21:00 GMT

November – Looking North

November – Looking North

Most of **Aquila** has now disappeared below the horizon, but two of the stars of the Summer Triangle, **Vega** in **Lyra** and **Deneb** in **Cygnus,** are still clearly visible in the west. The head of **Draco** is now low in the northwest and only a small portion of **Hercules** remains above the horizon. The southernmost stars of **Ursa Major** are now coming into view. The Milky Way arches overhead, with the denser star clouds in the west and the less heavily populated region through **Auriga** and **Monoceros** in the east. High overhead, **Cassiopeia** is near the zenith and **Cepheus** has swung round to the northwest, while Auriga is now high in the northeast. **Gemini,** with **Castor** and **Pollux,** is well clear of the eastern horizon, and even **Procyon** (α Canis Minoris) is just climbing into view almost due east.

At the beginning of the month (November 4) Daylight Saving Time comes to an end in North America.

Meteors

The **Northern Taurid** shower, which began in mid-October, reaches maximum – although with only a low rate of about five meteors per hour – on November 11–12, when the Moon is a waxing crescent, but gradually trails off, ending around December 10. There is an apparent 7-year periodicity in fireball activity, and 2018 is unlikely to be another peak year. Far more striking, however, are the **Leonids,** which have a short period of activity (November 5–30), with maximum on November 17–18. This shower is associated with Comet 55P/Tempel-Tuttle and has shown extraordinary activity on various occasions with many thousands of meteors per hour. High rates were seen in 1999, 2001 and 2002 (reaching about 3000 meteors per hour) but have fallen dramatically since then. The rate in 2018 is likely to be about 15 per hour. These meteors are the fastest shower meteors recorded (about 70 km per second) and often leave persistent trains. The shower is

very rich in faint meteors. In 2018, maximum is on days 10 and 11 of the lunation, with a waxing gibbous Moon, so conditions are not particularly favourable. Observing conditions will be best after midnight.

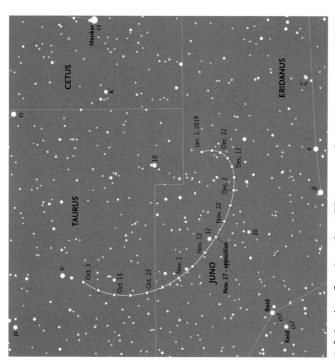

A finder chart for minor planet (3) Juno around opposition on November 17. Background stars are shown down to magnitude 8.5.

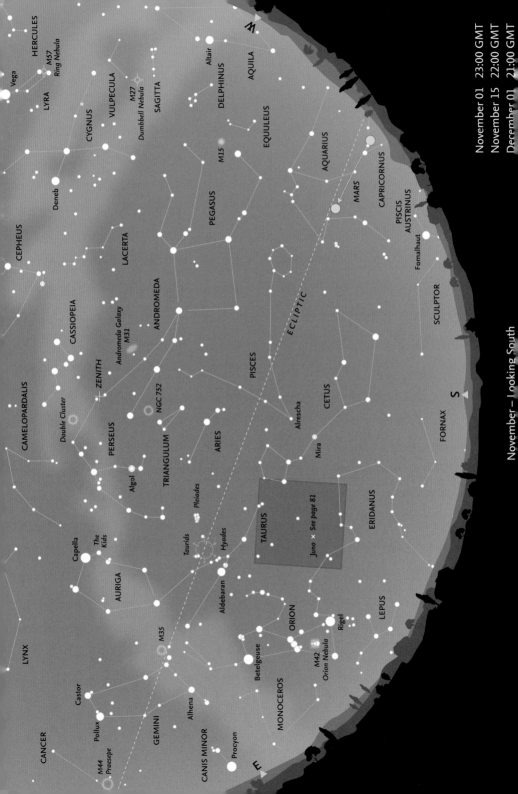

November 01 23:00 GMT
November 15 22:00 GMT
December 01 21:00 GMT

November – Looking South

November – Looking South

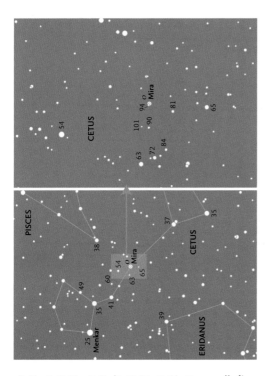

Orion has now risen above the eastern horizon, and part of the long, straggling constellation of **Eridanus** (which begins near **Rigel**) is visible to the west of Orion. Higher in the sky, **Taurus**, with the **Pleiades** cluster, and orange **Aldebaran** are now easy to observe. To their west, both **Pisces** and **Cetus** are close to the meridian. The famous long-period variable star, **Mira** (ο Ceti), with a typical range of magnitude 3.4 to 9.8, is favorably placed for observation. In the southwest, **Capricornus** has slipped below the horizon, but **Aquarius** remains visible. Even farther west, **Altair** may be seen early in the night, but most of **Aquila** has already disappeared from view. **Delphinus**, together with **Sagitta** and **Vulpecula** in the Milky Way, will soon vanish for another year. Both **Pegasus** and **Andromeda** are easy to see, and one of the lines of stars that make up Andromeda finishes close to the zenith, which is also close to one of the outlying stars of **Perseus**, high in the east.

Finder and comparison charts for Mira (ο Ceti). The chart on the left shows all stars brighter than magnitude 6.5. The chart on the right shows stars down to magnitude 10.0. The comparison star magnitudes are shown without the decimal point.

The Moon's phases for November

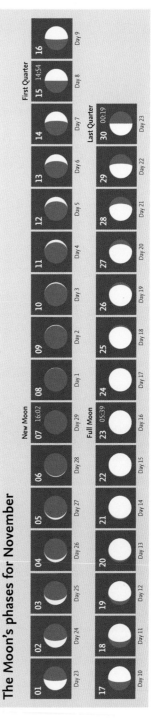

November – Moon and Planets

The Moon

On November 2 in the morning sky, the Moon is north of **Regulus** in **Leo**. On November 5, the waning crescent will pass north of **Spica** in **Virgo** and may just be glimpsed in the dawn sky. On November 16, the Moon (one day after First Quarter) is south of **Mars** in the evening western sky. The maximum of the Leonid meteor shower, November 17–18, occurs when the Moon is waxing gibbous, so is not particularly favourable. On November 23, the Full Moon passes just north of **Aldebaran** in **Taurus**. On November 29 it is near south of **Regulus** in **Leo** with closest approach occurring during daylight.

The planets

Mercury reaches greatest eastern elongation on November 6 (23.3°E, mag. -0.3) but is too low in the evening sky to be visible. **Venus** is in **Virgo**, but too close to the Sun to be visible. **Mars** moves from **Capricornus** into **Aquarius** and fades from mag. -0.6 to -0.1 over the month. **Jupiter** is close to the Sun and invisible in the dawn sky. **Saturn** (mag. 0.5) is still in **Sagittarius**, and may initially be glimpsed in the evening sky, but soon becomes too low. **Uranus** is still retrograding in **Aries** at magnitude 5.7 and **Neptune** in **Aquarius** at magnitude 7.9, retrogrades very slowly until it reaches a stationary point (on November 25) and then begins direct motion. The minor planet **(3)Juno** comes to opposition in **Eridanus** on November 17. (A chart is shown on page 81.)

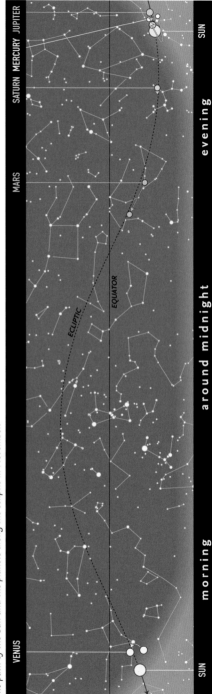

The path of the Sun and the planets along the ecliptic in November.

VENUS | SUN | morning | around midnight | evening | SUN

SATURN MERCURY JUPITER | MARS

ECLIPTIC | EQUATOR

Calendar for November

02	04:31	Regulus 2.1°S of Moon
04		Daylight Saving Time ends (N. America)
05–30		Leonid meteor shower
05	21:48	Spica 1.6°S of Moon
06	02:24	Venus 9.5°S of Moon
06	15:32	Mercury greatest elongation (23.3°E, mag. -0.3)
07	16:02	New Moon
08	17:36	Jupiter 3.8°S of Moon
09	11:11	Antares 8.6°S of Moon
09	11:35	Mercury 6.7°S of Moon
11–12		Northern Taurid shower maximum
11	15:32	Saturn 1.5°S of Moon
14	15:56	Moon at apogee (404,300 km)
14	23:14	Venus 0.2°S of Spica
15	14:54	First Quarter
16	04:17	Mars 1.0°N of Moon
17	21:39	Juno at opposition (mag. 7.4)
17–18		Leonid shower maximum
23	05:39	Full Moon
23	21:38	Aldebaran 1.7°S of Moon
26	12:12	Moon at perigee (366,600 km)
26	23:49	Pollux 7.1°N of Moon
27	09:15	Mercury inferior conjunction
28	28:00	Mercury 0.5°N of Jupiter
29	09:52	Regulus 2.4°S of Moon
30	00:19	Last Quarter

* These objects are close together for an extended period around this time.

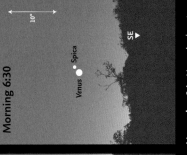

Morning 4:30

November 2 • The Moon with Regulus and Algieba.

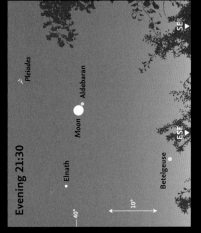

Evening 17:00 / **Morning 6:30**

November 11 • The Moon is close to Saturn. Nunki (σ Sgr) is nearby.

November 14 • Venus is close to Spica. There will be no conjunction.

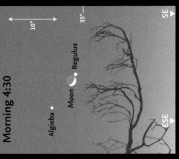

Evening 19:00

November 15–16 • The Moon passes Mars. Fomalhaut is close to the southern horizon.

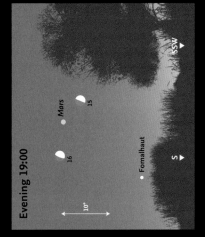

Evening 21:30

November 23 • The Moon is close to Aldebaran again.

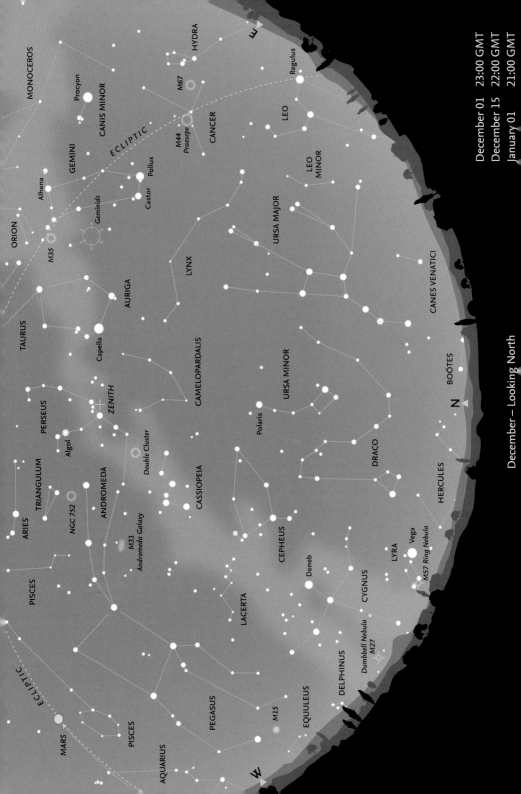

December 01   23:00 GMT
December 15   22:00 GMT
January 01   21:00 GMT

December – Looking North

December – Looking North

Ursa Major has now swung around and is starting to 'climb' in the east. The fainter stars in the southern part of the constellation are now fully in view. The other bear, **Ursa Minor**, 'hangs' below **Polaris** in the north. Directly above it is the faint constellation of **Camelopardalis**, with the other inconspicuous circumpolar constellation, **Lynx**, to its east. **Vega** (α Lyrae) is skimming the horizon in the northwest, but **Deneb** (α Cygni) and most of **Cygnus** remain visible farther west. In the east, **Regulus** (α Leonis) and the constellation of **Leo** are beginning to rise above the horizon. **Cancer** stands high in the east, with **Gemini** even higher in the sky. **Perseus** is at the zenith, with **Auriga** and **Capella** between it and Gemini. Because it is so high in the sky, now is a good time to examine the star clouds of the fainter portion of the Milky Way, between **Cassiopeia** in the west to Gemini and **Orion** in the east.

Meteors

There is one significant meteor shower in December (the last major shower of the year). This is the **Geminid** shower, which is visible over the period December 4–16 and comes to maximum on December 13–14 when the Moon is just before First Quarter. It is one of the most active showers of the year, and in some years is the most active, with a peak rate of around 100 meteors per hour. It is the one major shower that shows good activity before midnight. The meteors have been found to have a much higher density than other meteors (which are derived from cometary material). It was eventually established that the Geminids and the asteroid Phaethon had similar orbits. So the Geminids are assumed to consist of denser, rocky material. They are slower than most other meteors and often appear to last longer. The brightest often break up into numerous luminous fragments that follow similar paths across the sky. There is a second shower: the **Ursids**, active December 17–23, peaking on December 21–22, with rate at maximum

of 5–10, occasionally rising to 25 per hour. Maximum in 2018 is when the Moon is Full, so conditions are particularly unfavourable. The parent body is Comet 8P/Tuttle.

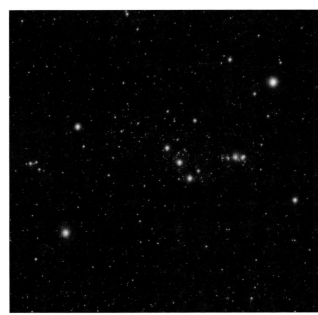

The constellation of Orion dominates the sky during this period of the year, and is a useful starting point for recognizing other constellations in the southern sky. Here, orange Betelgeuse, blue-white Rigel and the pinkish Orion Nebula are prominent. Orion can be found in the southern part of the sky (see next page).

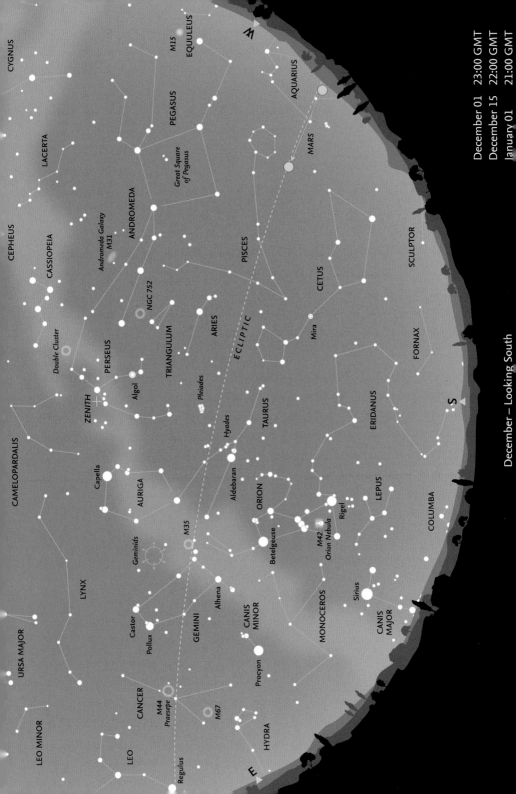

December 01   23:00 GMT
December 15   22:00 GMT
January 01   21:00 GMT

December – Looking South

December – Looking South

The fine open cluster of the *Pleiades* is due south around 22:00, with the *Hyades* cluster, *Aldebaran* and the rest of *Taurus* clearly visible to the east. *Auriga* (with *Capella*) and *Gemini* (with *Castor* and *Pollux*) are both well-placed for observation. *Orion* has made a welcome return to the winter sky, and both *Canis Minor* (with *Procyon*) and *Canis Major* (with *Sirius*, the brightest star in the sky) are now well above the horizon. The small, poorly known constellation of *Lepus* lies to the south of Orion. In the west, *Aquarius* has now disappeared, and *Cetus* is becoming lower, but *Pisces* is still easily seen, as are the constellations of *Aries*, *Triangulum* and *Andromeda* above it. The Great Square of *Pegasus* is starting to plunge down towards the western horizon, and because of its orientation on the sky appears more like a large diamond, standing on one point, than a square.

The constellation of Taurus contains two contrasting open clusters: the compact Pleiades, with its striking blue-white stars, and the more scattered, 'V'-shaped Hyades, which are much closer to us. Orange Aldebaran (α Tauri) is not related to the Hyades, but lies between it and the Earth.

The Moon's phases for December

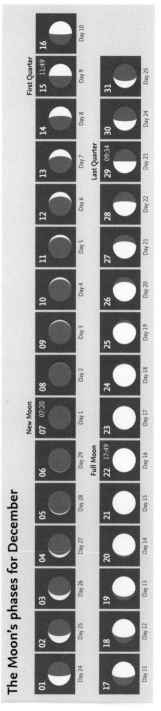

December – Moon and Planets

The Moon

On December 3, the waning crescent Moon passes about 8° north of *Spica* in *Virgo*, and will be visible in the early-morning sky. On December 9, the early crescent Moon passes 1.1°N of *Saturn* in *Sagittarius*. The planet may be visible just before it sets in the west that evening. On December 21, just before Full Moon, it is in *Taurus*, visible near *Aldebaran* in the evening. On December 30, one day after Last Quarter, the Moon is again in *Virgo* near *Spica*.

The planets

On December 15, *Mercury* reaches greatest elongation (21.3°W, mag. -0.5), and may be glimpsed in the early-morning sky. Early in the month, *Venus* is very bright (mag. -4.9) and in *Virgo*, but soon moves into *Libra*. It is visible in the east before dawn, ending the month at mag. -4.6, as it moves closer to the Sun. *Mars* begins the month in *Aquarius*, continues to move eastward and ends the month in *Pisces*, fading from mag. -0.0 to 0.4. *Jupiter* (mag. -1.7 to -1.8) initially in *Scorpius*, moves eastwards into *Ophiuchus*, becoming visible in the morning sky. *Saturn* is in *Sagittarius*, too low and close to the Sun to be visible. *Uranus*, at magnitude 5.7–5.8, continues to retrograde extremely slowly, beginning the month in *Aries*, and moves just over the border into *Pisces* by the end of the year. *Neptune* (magnitude 7.9), after reaching a stationary point on November 25, continues direct motion in *Aquarius*.

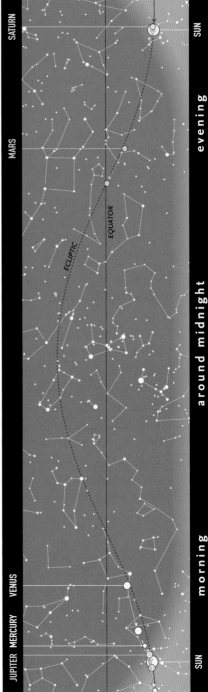

The path of the Sun and the planets along the ecliptic in December.

Calendar for December

03	04:15	Spica 7.7°S of Moon
03	18:42	Venus 3.6°S of Moon
C4–16		Geminid meteor shower
C5	21:07	Mercury 1.9°S of Moon
C6	13:23	Jupiter 3.5°S of Moon
C6	19:11	Antares 8.6°S of Moon
C7	07:20	New Moon
09	05:17	Saturn 1.1°S of Moon
12	12:25	Moon at apogee (405,200 km)
13–14		Geminid shower maximum
_4	23:22	Mars 3.6°N of Moon
_5	11:30	Mercury greatest elongation (21.3°W, mag. -0.5)
15	11:49	First Quarter
17–23		Ursid meteor shower
21	07:58	Aldebaran 1.7°S of Moon
21	15:00*	Mercury 0.9°N of Jupiter
21	22:23	Winter solstice
21–22		Ursid shower maximum
22	17:49	Full Moon
24	08:28	Pollux 7.0°N of Moon
24	09:49	Moon at perigee (361,100 km)
26	16:31	Regulus 2.5°S of Moon
29	09:34	Last Quarter
30	09:33	Spica 7.9°S of Moon

* These objects are close together for an extended period around this time.

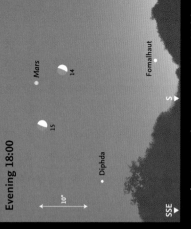

Morning 7:00

December 3–4 • The Moon passes Spica and Venus in the southeast, before sunrise.

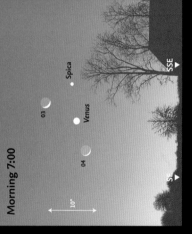

Early morning 4:00

December 21 • The Moon lines up with Aldebaran, the Pleiades and Bellatrix (γ Ori).

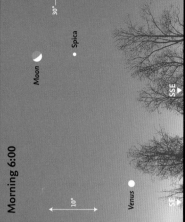

Evening 18:00

December 14–15 • The Moon passes Mars. Diphda (β Cet) and Fomalhaut (α PsA) are nearby.

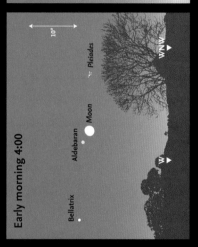

Morning 6:00

December 30 • The Moon with Spica. Venus is closer to the horizon.

Glossary and Tables

aphelion	The point on an orbit that is farthest from the Sun.
apogee	The point on its orbit at which the Moon is farthest from the Earth.
appulse	The apparently close approach of two celestial objects; two planets, or a planet and star.
astronomical unit	(AU) The mean distance of the Earth from the Sun, 149,597,870 km.
celestial equator	The great circle on the celestial sphere that is in the same plane as the Earth's equator.
celestial sphere	The apparent sphere surrounding the Earth on which all celestial bodies (stars, planets, etc.) seem to be located.
conjunction	The point in time when two celestial objects have the same celestial longitude. In the case of the Sun and a planet, superior conjunction occurs when the planet lies on the far side of the Sun (as seen from Earth). For Mercury and Venus, inferior conjuction occurs when they pass between the Sun and the Earth.
direct motion	Motion from west to east on the sky.
ecliptic	The apparent path of the Sun across the sky throughout the year. Also: the plane of the Earth's orbit in space.
elongation	The point at which an inferior planet has the greatest angular distance from the Sun, as seen from Earth.
equinox	The two points during the year when night and day have equal duration. Also: the points on the sky at which the ecliptic intersects the celestial equator. The vernal (spring) equinox is of particular importance in astronomy.
gibbous	The stage in the sequence of phases at which the illumination of a body lies between half and full. In the case of the Moon, the term is applied to phases between First Quarter and Full, and between Full and Last Quarter.
inferior planet	Either of the planets Mercury or Venus, which have orbits inside that of the Earth.
magnitude	The brightness of a star, planet or other celestial body. It is a logarithmic scale, where larger numbers indicate fainter brightness. A difference of 5 in magnitude indicates a difference of 100 in actual brightness, thus a first-magnitude star is 100 times as bright as one of sixth magnitude.
meridian	The great circle passing through the North and South Poles of a body and the observer's position; or the corresponding great circle on the celestial sphere that passes through the North and South Celestial Poles and also through the observer's zenith.
nadir	The point on the celestial sphere directly beneath the observer's feet, opposite the zenith.
occultation	The disappearance of one celestial body behind another, such as when stars or planets are hidden behind the Moon.
opposition	The point on a superior planet's orbit at which it is directly opposite the Sun in the sky.
perigee	The point on its orbit at which the Moon is closest to the Earth.
perihelion	The point on an orbit that is closest to the Sun.
retrograde motion	Motion from east to west on the sky.
superior planet	A planet that has an orbit outside that of the Earth.
vernal equinox	The point at which the Sun, in its apparent motion along the ecliptic, crosses the celestial equator from south to north. Also known as the First Point of Aries.
zenith	The point directly above the observer's head.
zodiac	A band, streching 8° on either side of the ecliptic, within which the Moon and planets appear to move. It consists of twelve equal areas, originally named after the constellation that once lay within it.

The Greek Alphabet

α	Alpha	ε	Epsilon	ι	Iota	ν	Nu	ρ	Rho	φ (φ)	Phi
β	Beta	ζ	Zeta	κ	Kappa	ξ	Xi	σ (ς)	Sigma	χ	Chi
γ	Gamma	η	Eta	λ	Lambda	ο	Omicron	τ	Tau	ψ	Psi
δ	Delta	θ (ϑ)	Theta	μ	Mu	π	Pi	υ	Upsilon	ω	Omega

The Constellations

There are 88 constellations covering the whole of the celestial sphere, but 24 of these in the southern hemisphere can never be seen (even in part) from the latitude of Britain and Ireland, so are omitted from this table. The names themselves are expressed in Latin, and the names of stars are frequently given by Greek letters followed by the genitive of the constellation name. The genitives and English names of the various constellations are included.

Name	Genitive	Abbr.	English name
Andromeda	Andromeda	And	Andromeda
Antlia	Antliae	Ant	Air Pump
Aquarius	Aquarii	Aqr	Water Bearer
Aquila	Aquilae	Aql	Eagle
Aries	Arietis	Ari	Ram
Auriga	Aurigae	Aur	Charioteer
Boötes	Boötis	Boo	Herdsman
Camelopardalis	Camelopardalis	Cam	Giraffe
Cancer	Cancri	Cnc	Crab
Canes Venatici	Canum Venaticorum	CVn	Hunting Dogs
Canis Major	Canis Majoris	CMa	Big Dog
Canis Minor	Canis Minoris	CMi	Little Dog
Capricornus	Capricorni	Cap	Sea Goat
Cassiopeia	Cassiopeiae	Cas	Cassiopeia
Centaurus	Centauri	Cen	Centaur
Cepheus	Cephei	Cep	Cepheus
Cetus	Ceti	Cet	Whale
Columba	Columbae	Col	Dove
Coma Berenices	Comae Berenices	Com	Berenice's Hair
Corona Australis	Coronae Australis	CrA	Southern Crown
Corona Borealis	Coronae Borealis	CrB	Northern Crown
Corvus	Corvi	Crv	Crow
Crater	Crateris	Crt	Cup
Cygnus	Cygni	Cyg	Swan
Delphinus	Delphini	Del	Dolphin
Draco	Draconis	Dra	Dragon
Equuleus	Equulei	Equ	Little Horse
Eridanus	Eridani	Eri	River Eridanus
Fornax	Fornacis	For	Furnace
Gemini	Geminorum	Gem	Twins
Hercules	Herculis	Her	Hercules
Hydra	Hydrae	Hya	Water Snake

Name	Genitive	Abbr.	English name
Lacerta	Lacertae	Lac	Lizard
Leo	Leonis	Leo	Lion
Leo Minor	Leonis Minoris	LMi	Little Lion
Lepus	Leporis	Lep	Hare
Libra	Librae	Lib	Scales
Lupus	Lupi	Lup	Wolf
Lynx	Lyncis	Lyn	Lynx
Lyra	Lyrae	Lyr	Lyre
Microscopium	Microscopii	Mic	Microscope
Monoceros	Monocerotis	Mon	Unicorn
Ophiuchus	Ophiuchi	Oph	Serpent Bearer
Orion	Orionis	Ori	Orion
Pegasus	Pegasi	Peg	Pegasus
Perseus	Persei	Per	Perseus
Pisces	Piscium	Psc	Fishes
Piscis Austrinus	Piscis Austrini	PsA	Southern Fish
Puppis	Puppis	Pup	Stern
Pyxis	Pyxidis	Pyx	Compass
Sagitta	Sagittae	Sge	Arrow
Sagittarius	Sagittarii	Sgr	Archer
Scorpius	Scorpii	Sco	Scorpion
Sculptor	Sculptoris	Scl	Sculptor
Scutum	Scuti	Sct	Shield
Serpens	Serpentis	Ser	Serpent
Sextans	Sextantis	Sex	Sextant
Taurus	Tauri	Tau	Bull
Triangulum	Trianguli	Tri	Triangle
Ursa Major	Ursae Majoris	UMa	Great Bear
Ursa Minor	Ursae Minoris	UMi	Lesser Bear
Vela	Velorum	Vel	Sails
Virgo	Virginis	Vir	Virgin
Vulpecula	Vulpeculae	Vul	Fox

Some common asterisms

Belt of Orion	δ, ε, and ζ Orionis
Big Dipper	α, β, γ, δ, ε, ζ, and η Ursae Majoris
Circlet	γ, θ, ι, λ, and κ Piscium
Guards (or Guardians)	β and γ Ursae Minoris
Head of Cetus	α, γ, ξ², μ, and λ Ceti
Head of Draco	β, γ, ξ, and ν Draconis
Head of Hydra	δ, ε, ζ, η, ρ, and σ Hydrae
Keystone	ε, ζ, η, and π Herculis
Kids	ε, ζ, and η Aurigae
Little Dipper	β, γ, η, ζ, ε, δ, and α Ursae Minoris
Lozenge	= Head of Draco
Milk Dipper	ζ, γ, σ, φ, and λ Sagittarii
Plough or Big Dipper	α, β, γ, δ, ε, ζ, and η Ursae Majoris
Pointers	α and β Ursae Majoris
Sickle	α, η, γ, ζ, μ, and ε Leonis
Square of Pegasus	α, β, and γ Pegasi with α Andromedae
Sword of Orion	θ and ι Orionis
Teapot	γ, ε, δ, λ, φ, σ, τ, and ζ Sagittarii
Wain (or Charles' Wain)	= Plough
Water Jar	γ, η, κ, and ζ Aquarii
Y of Aquarius	= Water Jar

Acknowledgements

D. Buczynski, Portmahomack, Ross-shire, Scotland: p.18, p.51

Adam Evans, CC by 2.0: p.21 (Andromeda Galaxy)

peresanz/Shutterstock: p.87 (Orion)

Steve Edberg, La Cañada, California: all other constellation photographs

Jens Hackmann: p.63 (Perseid meteor)

Damian Peach, Hamble, Hants.: p.13 (Comet Lovejoy)

Ken Sperber, California: p.69 (Double Cluster)

Editorial support was provided by Dhara Patel, Astronomer at the Royal Observatory Greenwich, part of Royal Museums Greenwich.

Further Information

Books

Bone, Neil (1993), *Observer's Handbook: Meteors,* George Philip, London & Sky Publ. Corp., Cambridge, Mass.

Cook, J., ed. (1999), *The Hatfield Photographic Lunar Atlas,* Springer-Verlag, New York

Dunlop, Storm (1999), *Wild Guide to the Night Sky,* HarperCollins, London

Dunlop, Storm (2012), *Practical Astronomy,* 3rd edn, Philip's, London

Dunlop, Storm, Rükl, Antonin & Tirion, Wil (2005), *Collins Atlas of the Night Sky,* HarperCollins, London

O'Meara, Stephen J. (2008), *Observing the Night Sky with Binoculars,* Cambridge University Press, Cambridge

Ridpath, Ian, ed. (2004), *Norton's Star Atlas,* 20th edn, Pi Press, New York

Ridpath, Ian, ed. (2003), *Oxford Dictionary of Astronomy,* 2nd edn, Oxford University Press, Oxford

Ridpath, Ian & Tirion, Wil (2004), *Collins Gem - Stars,* HarperCollins, London

Ridpath, Ian & Tirion, Wil (2011), *Collins Pocket Guide Stars and Planets,* 4th edn, HarperCollins, London

Ridpath, Ian & Tirion, Wil (2012), *Monthly Sky Guide,* 9th edn, Cambridge University Press

Rükl, Antonín (1990), *Hamlyn Atlas of the Moon,* Hamlyn, London & Astro Media Inc., Milwaukee

Rükl, Antonín (2004), *Atlas of the Moon,* Sky Publishing Corp., Cambridge, Mass.

Scagell, Robin (2000), *Philip's Stargazing with a Telescope,* George Philip, London

Tirion, Wil (2011), *Cambridge Star Atlas,* 4th edn, Cambridge University Press, Cambridge

Tirion, Wil & Sinnott, Roger (1999), *Sky Atlas 2000.0,* 2nd edn, Sky Publishing Corp., Cambridge, Mass. & Cambridge University Press, Cambridge

Journals

Astronomy, Astro Media Corp., 21027 Crossroads Circle, P.O. Box 1612, Waukesha, WI 53187-1612 USA. http://www.astronomy.com

Astronomy Now, Pole Star Publications, PO Box 175, Tonbridge, Kent TN10 4QX UK. http://www.astronomynow.com

Sky at Night Magazine, BBC publications, London. http://skyatnightmagazine.com

Sky & Telescope, Sky Publishing Corp., Cambridge, MA 02138-1200 USA. http://www.skyandtelescope.com/

Societies

British Astronomical Association, Burlington House, Piccadilly, London W1J 0DU. http://www.britastro.org/

The principal British organization for amateur astronomers (with some professional members), particularly for those interested in carrying out observational programmes. Its membership is, however, worldwide. It publishes fully refereed, scientific papers and other material in its well-regarded journal.

Federation of Astronomical Societies, Secretary: Ken Sheldon, Whitehaven, Maytree Road, Lower Moor, Pershore, Worcs. WR10 2NY. http://www.fedastro.org.uk/fas/

An organization that is able to provide contact information for local astronomical societies in the United Kingdom.

Royal Astronomical Society, Burlington House, Piccadilly, London W1J 0BQ. http://www.ras.org.uk/

The premier astronomical society, with membership primarily drawn from professionals and experienced amateurs. It has an exceptional library and is a designated centre for the retention of certain classes of astronomical data. Its publications are the standard medium for dissemination of astronomical research.

Society for Popular Astronomy, 36 Fairway, Keyworth, Nottingham NG12 5DU.
> http://www.popastro.com/
> A society for astronomical beginners of all ages, which concentrates on increasing members' understanding and enjoyment, but which does have some observational programmes. Its journal is entitled *Popular Astronomy*.

Software

Planetary, Stellar and Lunar Visibility, (Planetary and eclipse freeware): Alcyone Software, Germany.
> http://www.alcyone.de

Redshift, Redshift-Live. http://www.redshift-live.com/en/

Starry Night & Starry Night Pro, Sienna Software Inc., Toronto, Canada. http://www.starrynight.com

Internet sources

There are numerous sites with information about all aspects of astronomy, and all of those have numerous links. Although many amateur sites are excellent, treat any statements and data with caution. The sites listed below offer accurate information. Please note that the URLs may change. If so, use a good search engine, such as Google, to locate the information source.

Information

Astronomical data (inc. eclipses) HM Nautical Almanac Office: http://astro.ukho.gov.uk

Auroral information Michigan Tech: http://www.geo.mtu.edu/weather/aurora/

Comets JPL Solar System Dynamics: http://ssd.jpl.nasa.gov/

American Meteor Society: http://amsmeteors.org/

Deep-sky objects Saguaro Astronomy Club Database: http://www.virtualcolony.com/sac/

Eclipses: NASA Eclipse Page: http://eclipse.gsfc.nasa.gov/eclipse.html

Moon (inc. Atlas) Inconstant Moon: http://www.inconstantmoon.com/

Planets Planetary Fact Sheets: http://nssdc.gsfc.nasa.gov/planetary/planetfact.html

Satellites (inc. International Space Station)
> Heavens Above: http://www.heavens-above.com/
> Visual Satellite Observer: http://www.satobs.org/

Star Chart National Geographic Chart:
> http://www.nationalgeographic.com/features/97/stars/chart/index.html

What's Visible Skyhound: http://www.skyhound.com/sh/skyhound.html
> Skyview Cafe: http://www.skyviewcafe.com
> Sky & Telescope: http://www.skyandtelescope.com

Institutes and Organizations

European Space Agency: http://www.esa.int/

International Dark-Sky Association: http://www.darksky.org/

Jet Propulsion Laboratory: http://www.jpl.nasa.gov/

Lunar and Planetary Institute: http://www.lpi.usra.edu/

National Aeronautics and Space Administration: http://www.hq.nasa.gov/

Solar Data Analysis Center: http://umbra.gsfc.nasa.gov/

Space Telescope Science Institute: http://www.stsci.edu/